Of Castles and Colleges

Notes toward an Autobiography

The old brownstone castle in Rutherford, where it all started in 1942. Colonel Fairleigh S. Dickinson, who had a keen eye for aesthetic details, was responsible for the placing of the flagpole.

Of Castles and Colleges

Notes toward an Autobiography

Peter Sammartino

South Brunswick and New York: A. S. Barnes and Company
London: Thomas Yoseloff Ltd

A. S. Barnes and Co., Inc.
Cranbury, New Jersey 08512

Thomas Yoseloff Ltd
108 New Bond Street
London W1Y OQX, England

ISBN 0-498-01026-0
Printed in the United States of America

To my wife,

Sally,

who has always done half of the work
but who has rarely gotten any of the credit

Contents

Contents

Preface

"Peter, why don't you write a book?" was a phrase I have often heard over the years when I would recount informally some of the happenings at Fairleigh Dickinson University as we grew from 60 students in 1942 to 20,000 in 1967, when I retired as president. Then, one evening as we were leaving the St. Regis Hotel in New York, where a member of the Board of Fellows, Mr. Morris Leverton had been host at a small dinner party, one of the trustees, Mr. Samuel Silberman was to repeat the phrase. That very moment I made a decision. I would write a book weaving together the various events of the twenty-five years.

As I found a few moments from time to time, I would jot down some of the stories that I had been telling and put them in a file. I dreaded the moments when I would have to tie them together in some sort of order. I didn't want to write a history; that is for others to do. But I did want to recount some of the episodes that one remembers for one reason or another.

Education is a relatively simple process. Fundamentally, it concerns itself with the development of the mind. But we complicate this simple process with grandiose buildings, with complex administration, with the demands of vested interests, with inflexible curricula, and above all with an all-engulfing bigness, so that the individual involved is all but lost in the shuffle. But the individual is there and he demands to be heard. He is unimpressed with expensive structures at a time when taxpayers are going broke to provide them. He resents the unnecessary layers of administrative personnel. Family interests and restrictive courses of study are pushed aside. He seeks an antidote to the anonymity of the great horde in the tenderness of relations with fellow human beings, some of whom might be his teachers. His actions may seem abrasive and uncouth at times, but he is ever-ready for a person-to-person approach. Through the years, the things my colleagues (including my wife) found most important and most satisfying was this feeling of concern for the individual, and most of the stories I relate are involved in this concern.

I am greatly indebted to Professor Charles Angoff of our faculty, whom I consider one of the great writers of this century, for his patient criticism, and to Dean Lloyd Haberly, also of our faculty, and a great poet of our age, for coming to my rescue when memory would fail me. Mrs. Newton Foster was my savior in unearthing old photographs. My secretary, Mrs. Herbert Schuffenhauer, somehow deciphered my scrawls and worked out the puzzles of paragraphs that had to be inserted in outlandish places marked by long arrows that curled around pages.

I hope that the reader will find these weavings of interest. If he does not, let him blame the thoughtless people who egged me on to write this little book.

Peter Sammartino

Rutherford, New Jersey
May 27, 1971

Of Castles and Colleges

Notes toward an Autobiography

1

Two Martinis

Wouldn't it be fun to start a college in the old castle!

I was courting Sally Scaramelli, who lived on the corner of Montross and Fairview Avenues in Rutherford. I was on my second martini as we sat on the porch of the roomy house, looking at the old Ivison castle a half-block away. It was a gloomy old place, built in 1886 and now almost completely overgrown with trees and bushes. A caretaker lived in it, and three overgrown mastiffs discouraged any visitors. It was at that euphoric moment that Fairleigh Dickinson University was born, even though the actual incorporation didn't take place until eight years later. We were married that year and on weekends we would visit Sally's family. I was teaching at New College, which was an experimental college within Teachers' College of Columbia University. We were already doing at that time what many students are clamoring for today. Students and faculty met every Monday night to discuss the college and to work out solutions to problems.

The main thrust of what I have learned of higher education I got at New College, and particularly from the founder of that college, Dr. Thomas Alexander. New College was formed during the Great Depression as part of Teachers' College of Columbia University. In its ideals, its goals, its curriculum, it was to my mind far ahead of any college today. In those days we had neither the time nor the money to spend on public relations. Perhaps it was a mistake not to take time out and turn out publicity releases on what we were doing. Our main purpose was to train teachers who would not only know their own subject well but would have a strong cultural background.

Most important, we wanted these teachers to have a thorough understanding of their community and to be dynamic community leaders. As an example, when Rex Tugwell, then Under-Secretary of Agriculture, was working on the Resettlement Administration idea of establishing new towns, we at New College were collaborating on the idea of the teacher-leader in each of these communities.

Our students worked directly for the master's degree, picking up the bachelor's degree on their way simply as a traditional gesture. It has taken more than thirty-five years for this idea to start on its way as a standard for most teachers today. Let us remember that during the thirties there were still many teachers who were preparing for their profession in two-year courses.

All students had to spend a year abroad, and at one time New College, small as it was, probably had more students in Europe than all the rest of the undergraduate colleges put together. At that time "abroad" meant mostly England, France, Spain, and Italy. The students had to spend another period of time in a farm community and, as a matter of fact, we had our own farm community in Canton, North Carolina, which operated Spring to Autumn. It developed that a group of students could not afford to stay in college because of finances. It was suggested to them that they carry on their education in Canton through the year and in effect run their own community. We did have one faculty member to lead them and the rest of us would go down for short trips to give them help and teach seriatim. At other times we went over their reports by mail. I'll never forget the trip Sally and I took on the Thanksgiving weekend. I was teaching French and I thought it would be nice to add to their Thanksgiving fare by bringing down French onion soup, long loaves of French bread, and boxes of French pastry. We looked forward to succulent farm-grown turkey. Imagine our surprise when, after the steaming hot onion soup with warm French bread, the pièce de résistance was ground squirrel-meat patties! We were learning from the young people. This was the depression and they were living, believe it or not, on ten cents a day. They spent only for necessities and ate only what they could grow or trap. That first year was a nightmare for them but they stuck it through and with the careful garnering of their assets, by the next year they enjoyed beef, chicken, pork, and all sorts of vegetables and fruits. What a lesson they learned! Isn't this exactly what many of our young people are hankering for?

But the New College students had to do much more, too. They had to spend a period doing community and social work. We didn't want to train "ivory tower" teachers. We wanted them to know the problems of the sociological units in which we and other people live and they had to learn how to serve others. Isn't this exactly the idealism that inspires many of our young people today? Further, they had to spend a period in business or industry so that they would learn about this aspect of life. Finally, they were required to have a year's teaching internship. These were the out-of-class projects. Their course of study was equally different, exciting, and individualized. No two students followed the same courses or sequence. Each one, with the help of an adviser, mapped out a long-view plan and, within that plan, a shorter one for each semester. Students could take courses in any part of Teachers' College or of Columbia University. The whole college, faculty and students, met every week in the College Community Conference to take up the persistent problems of living. Another weekly meeting of the community discussed contemporary problems. Courses, discussions, seminars stemmed from these two integrative meetings. There was extreme flexibility, periodic assessment of results, and a willingness to experiment and to make quick decisions. What an inspiration for present-day America!

But to get back to the two martinis. One idea led to another and soon four of us from New College met on the same porch and with more martinis: John Taylor, who was later to become president of the University of Louisville; Paul Limbert, who as to become president of Springfield University; and Winifred Bain, who was to become president of Wheelock College. Some of our New College students aspired to, and did, become college teachers, but it was almost impossible for them, as new additions to a department, to try out any ideas they might have acquired at New College.

If we could set up a practicum college in the old castle we could give these young people a chance. Eventually the idea was interpolated as an item on the next New College budget. But this was the Depression decade and, as things got progressively worse, Columbia University cast out all new expenses. As a matter of fact, it reduced faculty salaries by twenty percent. The practicum idea was out. Many years later, I was to explain to President Grayson Kirk of Columbia how Fairleigh Dickinson University in New Jersey almost grew out of Columbia University in New York.

And so an idea was left dangling in the air—dangling until 1941.

In the fall of 1941, I gradually began to know many high school principals in northern New Jersey. One Sunday afternoon as my father-in-law, Louis J. Scaramelli, was tossing ideas back and forth, he said, "Why don't I invite some of the high school principals to a dinner, and we can discuss the idea of a new college?" In him the possibility of establishing a college struck a responsive note. He was a person who did not take America for granted. He was enormously grateful for the opportunity a new country had opened up to him. If he could play a part in encouraging the development of a new institution, it would be the greatest thing that could happen to him. He had had occasion to meet President Theodore Roosevelt and was tremendously inspired by him. One year, as a councilman in Rutherford, he was discussing the annual Memorial Day Parade. In going over the list of participating societies he noticed the name Ku Klux Klan. "Are they going to be masked?" he asked. When he was told they were, he said, "I believe the concept of the Ku Klux Klan is un-American. I can't believe they have the temerity to parade in our town. But if they parade masked, I shall personally tear the masks off their faces." In America, no one has the right to hide behind a mask! The Ku Klux Klan did not parade and was not heard from again.

Mr. Scaramelli believed strongly in education. Here was an opportunity to help repay his debt to America. He listened avidly to the things we had done at New College. And so we went ahead, drawing things a little closer by arranging for an organization meeting.

2

In the Beginning

"You must be crazy! Here we are wondering whether to close up for the duration of the war and you say that you're going to open up a college!"

Thus spoke the chairman of the New England group of college presidents assembled immediately after the general emergency meeting of all college heads held in Baltimore in January 1942.

To go back a little. The tentative decision to found a college was made in Rutherford on December 3, 1941. Present were sixteen high school principals, the assistant commissioner for higher education, Dr. Robert C. Morrison, Edward T. T. Williams, and my father-in-law. Four days later the Japanese bombed Pearl Harbor. What to do? I called an emergency meeting of the aforementioned sixteen high school principals who had constituted themselves a Board of Educaional Directors. It was this Board of a nonexistent college that elected me president. To tell the truth, I would have been disappointed if they hadn't. Their action was later accepted as a fact by the Board of Trustees when that body took legal form on March 16, 1942.

We discussed the alternatives of waiting for the end of the war or going ahead. The vote to go ahead squeaked through by exactly one. Soon after this vote the Federal Government called an emergency meeting of all college heads, but since Washington was so crowded the meeting was actually held in Baltimore. I heard of the meeting and I attended, even though I was a college president without a campus, without a student body, and without money.

After the general discussions on how to gear the institutions to war

17

needs, we broke up into sectional groups. I was so green that instead of attending the Middle States group in which New Jersey was included, I attended the New England group. As I remember there were about twenty of us present. At the request of the chairman, each one got up, gave his name and that of his college. When my turn came, I said, "I'm Peter Sammartino; we haven't decided on our name, but we expect to open a college in Rutherford in September."

And then came the unmerciful blast from the chairman that leads off this chapter: "You must be crazy! Here we are wondering whether to close up for the duration of the war and you say that you're going to open up a college!" I swallowed my pride and listened carefully to the discussion. I remember one Connecticut president's saying, "This emergency is going to give me a chance to bludgeon through the faculty some of the things I always wanted to do."

Sally and I went back to Rutherford somewhat crushed, but determined more than ever to go ahead. Actually, we had an advantage over the other institutions. We could start off with a war-geared institution. Of course, we had problems. First, we didn't even have a building. But we knew the old Ivison castle was empty and going begging.

The Ivision castle was an old landmark of Rutherford. It had been built in 1886 over the existing Tompkins homestead. As is the case with all such ventures, the owner, David Ivison, had tried to save money by utilizing the old structure. In the end it cost him much more to do it that way and would have been cheaper if he had razed the old building and started anew. Mr. Ivison, who had been a founder of the publishing company that was later to be known as the American Book Company, spent $350,000 for the pseudo-French chateau. Like many great estates, it had been abandoned by the family after fifteen years or so, and a partner of Burns Brothers Coal Company acquired it and added a wing with a swimming pool. This ownership lasted a few years and finally the building became the home of the Union Club, a gentlemen's club. Alas! the days of such clubs were over. During the Depression the mortgage payments were not met and the castle was taken over by the Rutherford National Bank, of which Colonel Dickinson was president.

Today the castle looks like a little place indeed. In those days it looked almost too big for us to handle. First we had to worry about the heating arrangements and luckily we were able to reactivate most of the old radiators. Then, to conserve heat, we closed up all the fireplaces. I was to repeat this and all of the other things when we did over Wroxton Abbey in England, but that is another story. The plumbing system had functioned for the Union Club and with a few changes and additions was put in order for students. We were so poor at this point that in one case, when I thought the tiling contractor wanted too much to replaster some tiles in a small toilet, I did the job myself to save sixteen dollars, I believe. The unprofessional character of my plastering can still be seen today in this particular room. We had eight contractors getting into each other's way, but somehow we got the floors sanded, the painting done, and the electrical wiring examined and replaced where necessary. We had to thin out a jungle of trees and leave the best to grow properly. We had no money for such fancy things as cement walks. Most of the office furniture and equipment was secondhand. We did splurge on new tablet-arm chairs for students. The kitchen was left pretty much the way it was, in hopes that neither of the two stoves would fall apart. I designed the chemistry laboratory and had a local carpenter build it for us for $200. When he finished he said, "Gee whiz, I never thought I could do it!"

Our first budget, that of June 6, 1942, was as follows:

Painting	$1300
Electricity	$ 200
Landscaping	$ 623
Plumbing	$1000
Carpentry	$ 800
Office Equip.	$1955
Furniture	$1000
Laboratory	$1000
Library	$3000

A word about Colonel Dickinson. He was a real gentleman of the old school, but a gay gentleman with a biting sense of humor. He had come to Jersey City from North Carolina on an old schooner as a cabin boy. The captain of the ship took a protective view of the young lad and saw to it that he got a job and went to night school. Later on in life the Colonel in turn took a protective view of the captain, established him in a comfortable home, and saw to it that his financial needs were met. Eventually Mr. Dickinson was to become an outstanding paper salesman, and he travelled all over the

United States with a great trunk full of samples. He earned $25,000 a year, a fantastic sum in the nineteenth century. One day in a drug store he met a man by the name of Maxwell W. Becton, who was selling thermometers. His entire sample case was barely larger than an étui and contained three or four thermometers. They engaged in conversation and discovered that they had both been born in North Carolina, barely fifty miles from each other. Mr. Becton expressed a desire to go into the thermometer business for himself but lacked the money. Out of that chance encounter came the establishment of Becton, Dickinson and Company, which in time was to become one of the great industrial empires of the world. Barely had the company been organized in New York City than the Spanish American War burst upon the scene. The new company soon outlived its quarters in Manhattan and moved to East Rutherford, New Jersey. The Colonel and Mr. Becton built homes opposite each other on Ridge Road in nearby Rutherford. Eventually Mr. Becton passed away. His house was acquired by the college and Sally and I still live in it.

The Colonel ran the company on old-fashioned principles. He was meticulously fair in all his dealings. If there was an argument about a bill, he would pay it quickly and ask the complaining party to look over the matter and see if perhaps there was an error. For years he walked two and one-half miles from his home in Rutherford to the factory in East Rutherford. The tower of the factory he placed so that it can be seen from the exact center of the main street of Rutherford— Park Avenue. He was precise in his speech. He was scrupulously fair to his employees and expected them to be equally fair in their production. He worked from an old fashioned rolltop desk in the corner of the wide office of his factory, from which he could oversee all the administrative and office staff. Nearby, Mr. Becton had his rolltop desk. Nobody had a private office—not even Ed Williams, who was his sales manager.

I got to know the Colonel when he was in his seventies. By that time he wasn't as spry as people tell me he was in his earlier days. But he still spent the morning and early afternoon at his company. In mid-afternoon he would stop by at the Rutherford National Bank. At that time the national banks could print their own bills, and the Colonel's signature appeared on his bank's bills, since he was its president. I always carry one of these bills in my wallet as a sort of nostalgic good-luck memento, but mostly because it reminds me of the Colonel.

Whenever I get into a conversation about the college, I usually pull out this bill with the remark that our University is the only one that has its name on money. And then I show them the signature "F(air-leigh) S. Dickinson." I'm told that in those early thirties it was possible to have some bills printed without the signature, which became legal tender when the president of the bank signed the bills in person. After all, couldn't a real signature be just as good as a duplicated one? At any rate, the Colonel is supposed to have had a few printed without his signature. Then when he might be in a restaurant with friends, he would pull out a twenty-dollar bill and give it to the waiter. In those days you could get a good meal for four people and still get change from a twenty-dollar bill. As the waiter would start to move away, he would call him back and say, to the utter consternation of his guests and the waiter, "Wait a minute, I forgot to sign that bill." At another time, he had attended a convention in Atlantic City and had had dinner in one of the great restaurants of the resort. He paid his bill with one of the Rutherford National Bank bills, this time fully signed. The cashier examined it with a quizzical look because, after all, it was slightly different from the usual bills. The Colonel told her not to worry and said "I make them myself. That's my signature on it." The girl laughed nervously and gave him his change. The Colonel got into his car and his chauffeur drove him away. After some miles, he heard the siren of the State Police. He stopped and the police car moved up. The officer got out and began to question the Colonel. He wanted to know about the bills that the Colonel was making. He was able to talk his way out of that one, but barely, while his wife, Grace, roared with laughter.

When the college was a legal entity and we started to fix up the castle in the early spring of 1942, the Colonel would always stop by after the bank visit and was always interested in the mechanical details involved. Sally and I were only too glad to have a surcease from our labors. Inevitably he would ask us to stop by his home at 185 Ridge Road in Rutherford. We would always have cocktails—Canadian Club with a little water and sugar.

My father-in-law knew the Colonel and was, in fact, a member of his bank board. I had asked him to approach the Colonel to ask for his financial help. The Colonel said, "I'll tell you what, Louis; I'll match whatever you give." This posed a dilemma. Mr. Scaramelli was in no position to give much. He was mainly responsible for the formation of the old East River bank in New York, had become a good friend of A. P. Giannini, the founder of the Bank of Italy in California (now the Bank of America) and had fought valiantly in the proxy fight when the Wall Street interests tried to force Giannini out of Transamerica. But my father-in-law had been virtually wiped out in the stock crash of the thirties. He had rebuilt his food-importing business, but, with the certainty of war, foreign trade was at low ebb, at least in his particular sphere. He could give only $15,000. I said I would add $15,000 and then the Colonel would have to match $30,000. Thus with $60,000 Fairleigh Dickinson was started. We made arrangements to buy the castle for $21,400. The next day the Colonel sent us a check for $25,000 and from time to time he would send us more checks, usually in amounts of about $10,000, to help us out. In terms of today's colleges, these sums were minuscule. At that time, they were sizable. For the first three years I drew no salary and three times I had to dig into my meager backlog to pay salaries to faculty and other personnel for the month.

But the help of the Colonel was not to be measured in terms of dollars alone. He began to like the idea of putting the old castle to worthwhile use. He took a genuine interest in what Sally and I were trying to do. And so did his partner, Maxwell W. Becton, who, in his quiet and gentle way, was always ready to encourage me and to help financially.

As the spring months went along, because I had a few gas tickets to use, the Colonel, his wife, Grace, and Sally and I would often go to the Colonel's favorite restaurant, the Grotta Azzurra, a cellar restaurant on Broome and Mulberry Streets in New York. We always had the same dish, broiled lobster with plenty of melted butter and Worcestershire, beautiful whole grain bread, fruit, and coffee. The Grotta has never had, and still doesn't have, cocktails. Our conversation would center usually on the mechanical details of our building in Rutherford. At that time Becton, Dickinson & Company was not a public company and was owned almost entirely by Colonel Dickinson and Mr. Becton. It was therefore perfectly possible for them to lend us a workman or two as construction problems arose. For instance, the walks we couldn't afford were built by Becton, Dickinson men. Our first library shelves were built by Becton, Dickinson carpenters and matched the selves in the Colonel's home on Ridge Road. Otherwise we would talk about some of the nostalgic things in the town. Later, as the first students started to register, we would talk about them and of their aspirations in life.

Our staff was small indeed and all of us slaved six days a week, day and night. Sally, without title and without pretensions, did everything that no one else could do or was prepared to do. Dean A. M. Wood, a gentle, hard-working soul who came to us from Nutley High School, did just about everything that I didn't have time for. The three of us could exchange tasks on a moment's notice. We had four full-time members of the faculty—one in English and literature, one in science subjects, one in psychology, and one in business subjects. All other courses were taught by part-time people, who lavished the same personal care on students as if they had been on full time. The librarian was one of the hardest working and most sincere persons I have ever known. Mrs. Althea Herald was a volunteer worker. How she was able to get our first library in working order, I shall never know.

The Colonel had a precise eye for architectural details. The first thing he did was to have a flagpole erected in front of the castle. It is perfectly placed and is still there. Then he gave us an organ. I needed a thousand things and a $10,000 organ was one thing I didn't need. All my life I have been the recipient of organs I don't need and can't afford to keep up. When we acquired our Madison campus we got an organ that had cost Mrs. Twombly $100,000, I am told. When Huntington Hartford gave us his gallery in Columbus Circle in New York City, it had an organ worth $75,000. The strange thing is that I am inordinately fond of the organ as an instrument and I love organ music. I hope some day to have a good organ of my own and to play it tolerably well. But an organ in a college is a dead expense unless you have an organ course.

In the meantime, James S. T. Ely, our volunteer lawyer, was helping us to get the college officially chartered in Trenton. We didn't want to call it Rutherford, because it singled one community out of many. "Boiling Springs," the old name of the area, was considered, but then we learned that there already existed a college by that name in South Carolina. The decision to use the name "Dickinson" wasn't made until the Spring of 1942, principally because the Colonel was against it. But the State Board of Education turned it down, fearing that Dickinson Junior College (we could start as a two-year college) might conflict with Dickinson College in Pennsylvania, one of the oldest colleges in America. And so in desperation we picked on the Colonel's first name, "Fairleigh," and came up with Fairleigh Dickinson. We practically had to bludgeon the Colonel into accepting the idea, but once the legal papers went through, he took quiet pride in having his name associated with the small but interesting educational development.

One must remember that in the early forties in northern New Jersey there were very few colleges. Paterson State Teachers' College was simply a series of classrooms in an old elementary school building. Bergen Junior College in Teaneck, which was to be merged with our institution, was a small group of frame buildings along the Hackensack River. Seton Hall College in South Orange was relatively small at the time and so was St. Peters in Jersey City. Montclair State Teachers' College and Jersey City State Teachers' College had what was for the period fairly good plants. The presidents of these two institutions, Dr. Harry Sprague and Dr. Allen W. Messler, respectively, were to join me at Fairleigh Dickinson and both were towers of strength when I badly needed help. And so was a later president at Jersey City, Forrest Irwin. I cite these details because it was human beings who have built the colleges of America. A retired college president freed from the burden of day-to-day operations can be of enormous help in an aspiring new institution and this is exactly what each of these men was.

I have the dubious distinction of having turned more old estates into colleges than any other person in the history of education. First, the Ivison castle in Rutherford, then the Peter Henderson homestead in Teaneck, although the work on this had already started before we took over. When I was ready to have a surcease, they really gave me a beauty when the Twombly Mansion was acquired in Florham Park. Then to top it all, the trustees decided to buy Wroxton Abbey in England. I didn't want the last two mansions, but once the board had made its decision it became my job to implement it.

When I think of how humbly Sally and I started in the old castle, I wonder how we had the energy to go on. As Jimmy Durante often says, "If I had to do it over again, I wouldn't have the energy." It was wartime, and one of the edicts was that we couldn't put an oil burner in the nineteenth-century boiler but had to use coal. It was next to impossible to get help, and when we did manage to get a janitor, he soon gave up trying to feed the furnace. The result was that, at times, between janitors, Dean Wood and I used to take turns heaving coal into the cavernous mouth of the furnace. On one such occasion, I had to rush back to my office to speak to Mr. and Mrs. Davis, whose daughter was one of our first students. They looked at me quizzically. It was very disconcerting but evidently I said the right things because their daughter, a bright youngster from Nutley High School, was to become an outstanding journalism student and wrote the words for our first college hymn. Later I found out that my face had been smudged with black dust from the furnace.

The students were involved in almost everything. One group, for instance, went out to secondhand-restaurant-equipment places on the Bowery and came back with almost new cafeteria furniture that was to last fourteen years. It was a good educational experience for them and it made the development of the college a part of their lives. Another group ran the little bookstore underneath the grand staircase during their spare hours. They ordered the books, kept inventory, and sold the books at specified hours. Another committee prepared the sandwiches for lunch and integrated their knowledge thus gained with the class on dietetics. These were the best sandwiches we have ever had at Fairleigh Dickinson University.

Every leak in the old castle was a minor tragedy. Roofer after roofer counseled us to put on a new asphalt tile roof to replace the real tile roof that was an integral part of the architecture of the old castle. Replacement tiles simply were not available. We didn't have the money for a new roof so we gamely protested that we didn't want to spoil the looks of the castle. Somehow we managed to patch things up and, curiously, twenty-eight years later, the old roof is still manfully protecting the old castle.

When it snowed, a farmer from Clifton came over with horse and plow and for the munificent sum of five dollars cleared out the walks from West Passaic and Montross Avenues to the Castle.

One day, Sally and I were looking out of the window of the castle and we saw two lonely cars in our tiny parking lot. I said to Sally, "Wouldn't it be nice if some day we had a few more cars?" I have learned not to joke about such matters. As we grew, the parking of thousands of students' cars turned out to be one of the nightmares of our existence. Periodically our neighbors have risen up in arms because the automobiles were blocking their driveways and jamming the streets.

Before we retired we were to see the Teaneck-Hackensack parking area for 9,000 cars. Whenever I took diplomats and presidents of a foreign university through our campuses, they were always amazed at the thousands of shiny cars in the parking lots. One ambassador said, "You've got more cars here than there are in my whole country!"

But in spite of our humble abode, it was an organic establishment and in future years I was to try to establish the same home-like atmosphere in Edward Williams College, which I shall describe later on. Today we spend untold millions for monumental buildings for students. Does it make the educational process any more vital? Do the grandiose buildings make the students any happier? In some cases they're blowing them up. Bigness and sumptuousness do not make better education. The best classroom is no classroom at all; the happiest classroom I have taught in was out on the green grass with the blue skies and the sun overhead. A classroom is a mechanical convenience to shield the students from the elements. I can truly say that those early days when we were ever so poor were the best and the most meaningful days of our existence.

3
Swans in Teaneck

"*And if there are* any brickbats, I'm all set for you."

This was when I, clad in a suit of armor, was speaking from the stage of our newly acquired campus in Teaneck to a group of about 150 rebellious and snarling students of what was formerly Bergen Junior College in February of 1954. I went on: "When you paid your tuition in 1953 it was to a bankrupt institution. We are here to help you receive an education. If you wish to stay, it will have to be on our conditions. If you don't like them, we shall respect your wishes and give you every possible help to transfer elsewhere. This help will be given to you without rancor or bitterness. However, we urge you to stay because this will be an opportunity for you to learn to participate in the building of a college. You'll have to work harder than you have been doing but we want you to succeed, and will give you every help to achieve your goal."

It is curious to realize that great and important things have always happened during the summer, when I mistakenly thought that I could do my work in a more relaxed manner in Maine. As a matter of fact, I have often told my Maine friends that our University really developed in Maine. It was in Yarmouth, Maine, that Sally and I mapped out the broad lines of the first campus in Rutherford. As a matter of fact, I was dividing my time between writing textbooks and drawing up ideas for a new college. It was while I was at New College that Mrs. Evelyn Vaill had suggested than I spend the summer at her place, Vaill Point in Yarmouth. I accepted and returned there for many summers. I had the habit of mentioning my friends in my

27

textbooks and of weaving their names into some of the lessons. And thus it is that on page 346 in my book *Avançons,* Evelyn Vaill is inextricably involved in a lesson on the French subjunctive. By 1941, the plans were fairly well fixed in my mind for an experimental new college. The Teaneck Campus was worked out in Harpswell, Maine, and the Madison Campus in East Sullivan, Maine.

At any rate, I was called in Maine to be told that the trustees of Bergen Junior College wanted to meet with ours to discuss a possible merger. It might be well to explain what happens when a private college goes out of existence. Its trustees cannot simply dispose of its assets and pocket the proceeds. If it is a nonsectarian institution, the assets go to the State. If it is a church-related institution, the assets revert to the church or order that established it. In a few cases a college, especially a two-year college, is owned by a person or group of persons. Usually these colleges are paying taxes and can be sold to other proprietors, or the land and buildings sold for profit. There are few cases left of the proprietary college, and most states, rightly, do not permit such colleges, New Jersey being a good example. Bergen Junior College had started as a proprietary college, but the new state laws had forced it to become nonprofit in character with a public board of trustees. As in the cases of many of the proprietary preparatory schools, when this happens, the old owner receives an annuity and perhaps a house to live in.

I rushed back from Maine and Mr. Hiram B. D. Blauvelt of the Bergen trustees, our chairman, Ed Williams, Dr. Hilleboe, Mary Mohair, Lee Moss, and I met at our home. Hi Blauvelt deserves a few words. Hi was one of the great sportsmen of New Jersey, if not of the world. He had been a Phi Beta Kappa student at Princeton and thereafter had studied for one year at Oxford. He took care of his family's commercial interests but this did not prevent him from being not a playboy but a bon viveur par excellence. He acquired some eleven hundred acres in Carlstadt, New Jersey, and proceeded to transform the area into one of the most exclusive shooting clubs in the country. Men came from far and wide to his pheasant shoots. He raised the pheasants himself at his estate, Bluefields, in Oradell, and on the day of the shoot thousands of them were brought to the central cages in Carlstadt. As I remember the shoots, the hunters were in sixteen-or-so stations in an outer circle. The birds were released a few score at a time from the center, but in various directions so that all guns would have an equal chance. Before the first group of birds was released, Hi would make the circle of the stations in a driven jeep in a Caesar-like fashion to receive the hurrahs of the participants. He always wore a tattered camel's-hair coat, even though he could afford the best.

On a few of these shoots I was privileged to ride beside him in complete awe and discomfort. On one occasion I brought along one of our university bagpipers and this added to the eeriness of the event and not a little to its corniness. Every so often, at a given signal, the gunners would move to the right one position, until each group had shot from each of the positions. Most birds didn't have a sporting chance. If one did get through, there was an outer ring of dead-shot employees who would bring it down. If he got through the second ring, there were always poachers in the swamplands beyond who managed to bring down the survivors. I am not a hunter, but I could not see the arrangement as a sporting one. After the shoot, there was a bacchanalian feast inside the clubhouse. Filled with the success of the carnage that had been inflicted on the luckless birds, the moments of triumph were duly celebrated with rivers of liquor, shouts of joy, excited horsing around, and loud singing. If you ever saw men thoroughly relaxed, thoroughly happy, and thoroughly carefree, this was it. The groaning board, which extended for perhaps twenty feet, bore up every type of food available, some of it prepared in the kitchen, much of it donated by the members. If a movie had been made of the eating that took place, most viewers would consider it exaggerated. Some of the viands, such as shrimp from Indonesia, were exotic indeed and might have been flown thousands of miles for the event.

Hi was the complete and undisputed master of ceremonies, and did he love the role of "le grand chef!" After the meal was over there was always the distribution of mementos. Most of the men present were very wealthy indeed, and it is amazing how each one would reach out avidly for the trinkets that were passed out. I remember, on one occasion, the souvenirs included bull's nose rings, a pocketknife, and an ashtray. The afternoon wore on less noisily as the stupor brought on by the alcohol and the food diminished the exuberance of the hunters. People began to leave, and it was the unwritten rule that you bought one or a dozen braces of the slaughtered birds to help defray the expenses of the club.

Hi never wore a tie—never. At that time we still had clothing standards at the college. It was embarrassing to us to have a trustee on campus without a tie when all students were expected to wear one. Today this sounds archaic, but customs have been changing over the years. I finally solved the problem by telling Hi I was having some special Fairleigh Dickinson bow ties made which he could easily clip on every time he attended meetings or was on campus. He liked that, even though he put on a show of pouting and suffering resignation in acceding to my request for trustee dignity.

The Junior College had been organized in 1933 and during the veteran influx had reached a peak enrollment of 1,400 students. It gradually lost students until it had about 300 matriculated full-time students. It had been unable to secure regional accreditation, and this in turn produced the downward spiral. The less money taken in, the less there was to spend on needed improvements and student services. The institution had reached the point where it had to sell off pieces of its property in order to pay faculty salaries. The trustees were all public-spirited men who were ready to try anything to reverse the decline. But unless you can find a generous amount of money and a knowledgeable and energetic administrator, it is difficult to save an institution that is foundering. A word about trustees. I have known many from many private colleges. They are all conscientious and desire to serve their institutions. Some of them don't realize that as trustees they are supposed to contribute and to help raise funds, and this is especially true today. A trustee has a difficult position. If he involves himself too much in the activities of the college, he may be accused of interfering. If he does nothing but sit passively at meetings, he is charged with being too inactive. At Bergen Junior College, a fund-raising campaign had not produced very much. Dr. Kron, who had assumed the chairmanship in 1952, tried valiantly to bring the institution around to a favorable position. Finally he felt that the best procedure would be to seek a merger with a larger institution. The choice was between Rutgers and Fairleigh Dickinson. For various reasons it was decided to merge with our own institution. The trustees therefore voted to terminate Bergen Junior College as such and to have Fairleigh Dickinson take over its assets and liabilities.

Our original campus is in Rutherford, roughly nine miles from Teaneck. By the fall of 1953 it was literally bursting at the seams. It had over 1200 day students and almost 1800 evening students, and was the third largest collegiate institution in the State. The diminution of veteran enrollment had had no effect on our overall registration. Overnight we found ourselves with a new campus which we were agreeing to develop as the Teaneck campus of Fairleigh Dickinson College on an equal basis with the Rutherford campus. Just as there are mergers in industry there are mergers in education. And within the next decade or two, there may be a lot more.

One of my hobbies during the forties had been collecting armor. Since the Teaneck incident, I have sent to our campus in England, Wroxton Abbey, whatever pieces I had, and there they may be seen to this very day. At any rate, taking over a small moribund college and making it into a new co-equal campus was a new experience for me, as it would be for most college administrators. It's easier to start from scratch than it is to attempt to use the imperfect and dissident components of an existing institution that resents being taken over.

In meeting with the Bergen Junior College students, I felt I had to get their attention even if I was going to be clobbered. As a pure gimmick and also because I had always wanted to wear one of those d———— suits of armor, with tongue in cheek I made my appearance before a startled student body. As I spoke it slowly began to dawn upon the students that Fairleigh Dickinson College was really their savior. I gave them a simplified accounting of the college budget for the Spring semester. I also gave them a full report on the merger and explained to them that their junior college has two alternatives open: to fade out of existence altogether or to cease as Bergen Junior College and to let another institution take over the educational activities. I invited them to choose a committee that would work with us on the budget and on the rebuilding of the campus. After all, you couldn't blame these distraught young people. They had seen us drop some of their comrades because of their poor records and some of them felt insecure enough to feel the cold steel of the administrative axe not far from their necks. Furthermore, in Rutherford we did not permit secret societies. The same rule was invoked for the new campus. This did not make friends in the fraternities that were running everything. The juke box room seemed to attract most of the students most of the time. We threw out the juke box. The library rarely had any students. The boys and girls were having a ball. Now it was all over.

But the merger meant that we could equalize the enrollments and organize two small colleges which would make better sense than one institution too large for its comfort and one too small for efficient operation. The first step was the physical rehabilitation of the buildings and facilities of the new campus. Anybody who thinks that such problems have nothing to do with real education is being extremely unrealistic. The decision to merge was made in mid-November. Almost overnight everything had to be done.

With various groups of faculty, we examined everything on the campus during two whole days—and I mean whole days. We would start out at eight in the morning, go right through to six, take an hour out for dinner, then work through the evening. It is amazing how much one can accomplish by working around the clock. Within 48 hours, we had made up a list of 24 physical projects. The next day, the sign-maker made a large sign listing the projects with horizontal lines next to each. Every week the proportion of the project completed was marked off in red so that all faculty, students, office staff, and custodians could see the progress being made.

The Teaneck campus was deficient in good classroom space. Within two weeks we had the preliminary plans for a fourteen-classroom building to bring up the complement of available classrooms to thirty-four. An enlarged students' commons had to be provided. The laboratories were in a mess. We had to rebuild them and order needed equipment. Offices were strewn all over the campus. We brought them all under one roof. Into one central building were placed admissions, registration, guidance, administrative, and financial offices. The faculty conference and lounge rooms were also put in the same building.

Every classroom and every office contained from six to fourteen different styles of furniture, all in various stages of disrepair. We threw out the worst of the lot; we sorted the rest and set a team of carpenters and refinishers to work on getting the better pieces into first-class condition. And, of course, we bought a great many new things. And the junk that had accumulated! I almost invited my friend Helen Worden Erskine, who has made a hobby of studying the habits of acquisitive misers. Old beds, faded promotion literature, cast-off living room furniture, rusted metal cabinets, empty bottles, patched-up lamps, old magazines, stacks of term papers! But this is what happens in most institutions. Unless someone physically goes through each building periodically, educational institutions become catch-alls for all the debris of decades, which adds expense but not one iota of education for the students.

But the greatest surprise of all was stumbling into a room filled with what were at one time good business machines just thrown about pell mell. If they had been in use, they could have been sold for foreign export in the postwar period. Again we salvaged what we could and put the machines into prime condition—about $15,000 worth, including one that typed electrically individual copies of a form letter.

In order to have industrial leaders, high school teachers, and students know the new campus, we had to arrange for a long series of dinners, teas, and luncheons, and that meant that standard food service procedures had to be set up and special equipment had to be provided.

Not the least of the problems was the stream that meandered through the most important part of the campus. When it was dry it was an unsightly ditch. When it was running it created an unhealthy swamp overrunning the campus. We decided to fill in the low areas, and to insure a running stream at all times by building a dam and creating a pond which could reflect the supporting pillars of the new classroom building. On the pond I placed two swans that were a gift from my friend Dick Poor, and hence the swan on our coat-of-arms. Isn't corn wonderful!

The library had to be brought up to four-year standards, and it had to be done overnight. Within a week, all the book orders were made up and sent out. Each course was sure to have the reference books it needed in time. Two wings in the library building that were being little used became ideal classrooms for seminar study. Again our librarian angel, Althea Herald, rolled up her sleeves, threw out the worthless books, and built up the library. She built so fast that twice in a decade we had to build a new library for her.

The objective evaluation of instructors became what was probably the most pressing and the most important human problem. On the one hand, there was a great moral obligation to be as fair and as considerate as possible to the faculty of the institution that had been absorbed. On the other hand, there was an equally strong need not to add teachers who might not fulfill the needs of the larger institution.

As it turned out, only three of the full-time instructors had to be dropped. The other five became enthusiastic and hardworking members of our own faculty. I have nothing but pleasant memories of them as lifelong friends and professional leaders.

Even the integration of the bookstores had a basic educational importance. It is extremely necessary to have all the books needed for courses ready for distribution at the beginning of the semester and to make sure that the students acquire them quickly during the first few days of classes. Effecting a merger in midyear meant that, wherever possible, we had to try to avoid extra book expense by the students in the absorbed institution. Naturally, the books would not be the same even in similar courses in the two institutions. Thousands of books had to be sent back to publishers; thousands had to be ordered in a hurry.

The people who reconstituted the science laboratories were a joy to behold. Kathleen Hillers literally went through the locked-up caches of chemicals of a half dozen previous instructors, each one of whom, rather than bother with the stock of his predecessor, would just order a new stock. She must have salvaged $20,000 worth of chemicals. Julius Luck sorted out the physics equipment that was actually heaped in a storage room. He salvaged, he ordered, he built, and in two months he had worked a miracle. Ray Miller, our Dean, spared no hours in helping me to coordinate everything and I'm sure that the Teaneck experience must have been invaluable on the splendid job he did in building York College in Pennsylvania, of which he became president. Sally got out a new catalogue in less than a month. Clair Black, Loyd Haberly, and Harold Feldman coordinated the curricula in short order. Olive Foster singlehandedly did what a corps of Madison Avenue public relations writers would have charged oodles for, and sent out the reams of releases. No one person got a cent extra for building a new college practically overnight. It was all part of their professional responsibility and it was a beautiful operation of coordination, of unwasted effort, of creative exhilaration. Each person seemed to know exactly what to do. Those few members of the Bergen staff who were left got into the spirit of the changeover and collaborated fully, and became enthusiastic members of the team. When it was all over, I called the Rutherford faculty who had built the new campus and, before the assembled students, presented each with a bouquet of roses. This was their sole reward. In the midst of things, I received a telephone call from Al Driscoll, then Governor of New Jersey, telling me how happy he was that we had taken over and wishing us luck. It was just the sort of encouragement I needed and I told him many times later how important his call was. Incidentally, a few years later, he became a member of the newly-formed Board of Fellows.

Of one thing I felt proud. The phalanx of instructors and administrative officers that carried through the integration quickly and expeditiously was a joy to behold. And they too experienced a thrill in the professional challenge involved. Just as years ago I experienced a thrill in participating in the creation of New College at Columbia University, so these men and women were enjoying helping to create a new collegiate center. They knew the philosophic framework of the college. They knew the standards they were aiming for. They knew the importance of the physical plant and of the mechanics involved in the everyday operation of a college. They knew the techniques of student guidance: bold leadership when necessary, democratic participation wherever possible. They knew above all that while the task seemed almost impossible at first, if something was accomplished every day the mountain would soon be whittled down. They enjoyed the freedom of being able to make quick decisions whenever it was necessary to do so. They knew full well that not all decisions were going to be correct, that the best one can do is to act intelligently on the basis of the facts at hand. And now we were embarked on a new experiment in community education—one college with two campuses within nine miles of each other. It would be fun to see how they would develop. I had been saying, "If I had to do it all over again, I wouldn't have the energy." I've learned since not to joke about such matters. We *had* to do it all over again!

As far as I have been able to find out, our Teaneck campus was built faster than any college I know of. Although the pourparler had started in the summer of 1953, the lawyers had dilly-dallied for months. In mid-November, the till was empty and there was no money to pay the faculty. One trustee told me that some of them were so concerned that they would be personally liable for debts incurred. At this point the pressure was on to rush through the formal papers of merger. We had a little over two months to get the new campus ready for the February opening and a little over nine months for the fully completed buildings. We moved no less than seven wooden buildings; roads had to be built; marshes had to be drained. In Rutherford we had had so much trouble with parking that I vowed that we would always have several hundred more parking places than we needed. Because of lack of money, no painting or maintenance of property had been done for five years at Teaneck. Every square inch of property had to be painted and put in prime condition. Simultaneously, curricula had to be established, guidance procedures set up, and student activities generated on a rational basis.

Some curious things did happen. One of the old administrators would spend most of the afternoon sleeping in a cozy room in the old homestead. Another one literally locked himself in his office, apparently afraid to face students or adults. Luckily, one soon retired and the other went elsewhere. One instructor tried to incite a clique of students to open rebellion. I believe that the only reason she did not succeed was that we were determined to give students the best instruction and the best guidance we could. An example of what we did was in the art department, which had the largest registration. I had an outside consultant come in, Stanley Lapolla, who later was in charge of art for New York City, and give each student as much time as he or she needed to think out educational goals. Many of them, with at best very mediocre art talents, were majoring in art, hoping to make a career in what is probably the most competitive field of all. It was a pure waste of time and money. A few did have talent, but it was much better for them to go to a professional art school, which we were not. Stanley, who served us ever so diligently, refused to take a single cent for his services. When he sat down with a student he gave his all, and had only one aim, to help that person to achieve the best cultural and vocational goals. Mind you, he was not trying to build up for any vested interest nor was he trying to build up enrollment for the college; he was trying to help a human being, often the very thing we forget in colleges.

We had to build a cohesiveness, an élan, a singleness of purpose. I can think of no better way to do this than to have everybody working away. Every once in a while, sometimes twice in one week, sometimes not for a month, we would all get together, evaluate what we were doing, discuss students' problems, and decide on corrections in our course. We met with students for two reasons: to keep them informed on what was happening and to get their point of view. Not much later, I was faced with an attempted unionization of our custodial staff. I called in the student council and gave them a complete picture of the situation. I pointed out the union organizer and explained to them how he works. I told them that the college was not against unions and respected the men's wishes in this matter. I told them, however, that the problem was not that simple. Our custodians filled in the summer months by doing odd jobs to keep up the physical condition of the buildings, much as their own fathers might keep up their houses with painting, gardening, and minor repairs. Unionization would limit the scope of activities. Most custodians would have to be laid off during the slack summer months and specific union trade brought in. I explained to the students that I had asked the advice of two outstanding union leaders who were friends of mine and could advise me behind the scenes. They made independent studies quickly of competing scales for school custodians. They examined the whole situation and then reported back to me. They advised me against unionization for three reasons. First, our own men would earn less year round under the arrangement. Second, unionization would add immeasurably to administrative procedures with increasing nonproductive costs that would have to be borne by students. Third, the leadership of the particular union had an unsavory reputation. They did recommend one fringe benefit we did not have: funeral cost allowance in case of death. There are times when a president has to make a quick decision even if he doesn't have clear authority to do so. I made the decision to have the allowance and immediately explained to the trustees why it was necessary to maintain a competitive position. The students and I met for about one and one-half hours in an atmosphere of complete candor. At the end, I asked for a closed-ballot vote as to whether we should agree to a union or not. Unanimously, they voted against a union. The custodians, in the meantime, obeying the rough tactics of the organizer, had set up a picket line. Since he couldn't get enough of our own men, he brought in outside pickets. The students decided to set up their own cleaning schedules. We explained everything to our own custodians and told them why we felt a union was inadvisable. Most of them merely stayed home with unfortunate deductions in salary. At the end of two weeks, neither the organizer nor his outside pickets showed up. The men with great relief resumed their jobs. We provided as much overtime as we could to make up for lost wages. The students had had a vivid lesson in unionization, and let us not forget that the fathers of an appreciable number of the students were union members themselves. While the strike was going on, John Eisenhower and his wife, Barbara, visited the Teaneck campus. Colonel Eisenhower had accepted a position as provost of the Teaneck campus. I explained to him exactly what was happening and I am glad to say that the matter made no difference in his decision to come with us. Six months later he did have to withdraw his acceptance because of events at the White House. A month or so after that, I was guest at a White House dinner. As I entered the White House, the man who greeted me at the door was a Fairleigh Dickinson graduate. President Eisenhower greeted us a few minutes later and told me how much he enjoyed reading my book, *The President of a Small College.* Fairleigh Dickinson was a small college then, and I had written a little book pointing out some of the problems of a small institution. John Eisenhower had in the meantime become a military aide at the White House. He showed me around the various historic rooms and then we returned to the informal discussion with fifteen other college presidents on some of the problems of contemporary institutions.

Within a year we had all but forgotten that Teaneck was foreign territory. It was now Fairleigh Dickinson, fully infused with new money, new personnel, new ideas. The enrollment quadrupled as the high school principals and guidance counselors understood what we were trying to do. As we built up the laboratories and the library, they realized the changes that had taken place and recommended their graduates to the new campus.

But as we built the Teaneck campus, interesting things happened. They had no lasting effect but they served to lighten the tremendous responsibility of building a new college. Our student cafeteria was very small indeed. The meager capacity was made even smaller by the fact that if two students each with a standard tray sat down at a table for four, the trays prevented others from using the table. I quickly thought of the idea of having triangular trays, four of which would fit neatly on the square, bridge-table-size tables. That week I was having by portrait painted by Tadé Styka in his beautiful studio on Central Park South. I was watching his hard-carved palette and suddenly I got an idea. Why not have a hole in the triangular tray so that the student could rest it on his arm and poke his thumb through the hole and still carry a briefcase in his right hand? Within two weeks we had manufactured our first Tray-Angles for the cafeteria. I registered the name, applied for a patent, and assigned any future profits, in case they were commercialized, to the scholarship fund of the college.

The estate at Teaneck had been originally created by Peter Henderson, whom many will remember as the "Seed Man." He had all kinds of trees planted on the estate, including two cedars of Lebanon. It was natural that these trees attracted many different kinds of birds. With the help of the Audubon Society I created a bird walk, which went along the Hackensack. I wish I could say that the idea was successful. A few ladies and a few students would take the walk, but as each new building went up, the bird sanctuary gradually disappeared, as it had at Rutherford where I tried to do the same thing and where one may see in front of the castle a birdbath I used to fill myself whenever the single custodian of the early days forgot. In the pond I created, I put in along with the swans a dozen ducks and made sure that they were fed. Alas, they proved too much of a temptation for the local gourmet, who liked nothing better than a good duck dinner. I also had bags of feed thrown on the banks of the Hackensack, which washed our property, and this did attract hundreds of ducks. Unfortunately, it also attracted hunters on the opposite shore in Hackensack, who had a shooting preserve all set up for them in their own backyard. What made it still worse was that some of them were poor shots and we periodically found bullet holes in our academic buildings. It was a disappointing experience, but perhaps we expected too much in a rapidly growing area.

Soon after we acquired our Teaneck campus I realized that we would need more land. I knew the owner of the adjoining property, Davenport West, and I went to see him and asked him to consider selling his property to us. He told me that his home was a family homestead and that he enjoyed living in it. No, the college would have to get along without the Phelps property. A few years passed and Davy came to see me. He had seen building after building going up and he realized he was too close to progress. As a matter of fact, our property flanked him north and south and there was a continual parade of students across his sidewalk. We arranged quickly for a sale, which made me happy not only because we needed the intervening property but because I realized how much of a nuisance we must have been to his idyllic estate. I am glad to say that he bought another beautiful property in nearby Englewood that was in no danger from encroaching educational institutions.

4

Nutrition and the College

"Peter, your girls look very pretty but they are living on only 50 per cent health. Why don't you do something to have them enjoy optimum health?"

It was this remark by Edward R. Hewitt that kindled my interest in nutrition. Mr. Hewitt, who lived in an old-fashioned brownstone in Gramercy Park, had written his autobiography, *Those Were the Days.* It was a most fascinating account of the early years of the century. I invited Mr. Hewitt to speak to our students in the spring of 1942, a war year when practically all of our students were girls.

Mr. Hewitt was one of the most scientifically minded men America has ever had. He and his brother had been involved in almost every major mechanical and scientific innovation of the century, whether it was the Hewitt arc light, the Hewitt motor, or the iron works in Trenton. The only other person in history with whom I can compare him in interest in inventions is probably Edison.

Hewitt had been asked by a friend to find a use for the soya bean. He first determined that it made an excellent protein feed for hogs. Then he went a step further and began to experiment with it as a good food for human beings. This led to his researches in vitamins. In his old-fashioned brownstone he set up a little laboratory. Soon he was helping his friends in nutritional problems and in overcoming vitamin deficiency. He advised me to become interested in vitamins. But it was wartime and I had other things on my mind.

But some years later, Dr. Roy Dufford Ribble—who had been my dentist and in whose office at Rockefeller Center in New York I had broached the subject of establishing a school of dental hygiene—decided to cut down his practice and come to the College to set up the school. One day Dr. Ribble invited a Dr. Fred Miller of Altoona, Pennsylvania, to lecture to the students on dental health and nutrition. I remember that, although I was very busy, I went to the lecture as a courtesy to Roy. But as I listened, I became more and more fascinated with what Dr. Miller was saying. By the time he was through, I had made a decision. I was going to encourage students to enjoy better health. This is a good example of how far-reaching decisions are made quickly. I made a survey of students' eating habits. They were horrible. Most students ate little or no breakfast and if they did it was the wrong kind. Naturally, about ten-thirty or so, their blood sugar low, they would crave food. Usually it was a bar or two of candy, which would shoot up their blood sugar level for twenty minutes or so, giving them an illusion of quick energy. But soon after, the blood sugar level would drop lower than before. What to do?

We went on an all-out campaign to encourage students to have a full breakfast—fruit, eggs or meat, whole wheat bread, milk. This would keep their energy level up till lunch time when a protein, vegetable, fruit, and milk would keep them going until the late afternoon. The college featured full breakfasts at the lowest prices in its cafeteria. Still there were students who couldn't be bothered with breakfast. I had come to know Adelle Davis, the popular writer on nutrition, and knew of her "Tiger's Milk," which was really the same basic idea that was later included in Metrecal. At that time, Adelle encouraged people to make their Tiger's Milk with dry brewer's yeast, powdered milk, and fruit juice. We told the students that if they had no time for breakfast, to gulp down a glass of Tiger's Milk, which we supplied free of charge. The most drastic decision was to discontinue commercial candy and carbonated beverages in our cafeteria. The students made their own candy, using honey, powdered milk, and peanut butter, plain or with nuts or raisins.

I remember calling them into my office and as I was talking I gave a bowl to one of the students. Then I gave her peanut butter to spoon in. I gave another student a box of powdered milk and told her to throw some in the bowl. I gave a third student some honey and told her to pour some onto the other ingredients in the bowl. By this time the students thought I was somewhat daffy. "Now mix everything up!" I said. Then I gave her some raisins to mix in. As the mixture took on a taffy-like texture, it began to dawn upon the students what I wanted to show them. "Now taste it!" I said. They did, and liked it. I asked them to make it for their fellow students. The college was small enough so that a small committee took on this task with alacrity. This didn't make the cafeteria manager happy, for his operation wasn't making a profit on this kind of candy. But at least this was a sweet that was highly nutritious and contained lasting proteins. We took out the carbonated beverages because they substituted sugar and chemicals for real nutrition. We told students that if they were thirsty they could have milk, fruit juices, or water. At that time there were no sugar-free carbonated beverages, which were to be more injurious to good health than those made with refined sugar. Then we began to make our own bread. Mrs. Rudkin, who started the Pepperidge Farm bread, helped us. By this time, Dr. Clive McCay of Cornell had become interested in what we were doing and both he and Mrs. McCay encouraged us and helped us tremendously.

My friendship with Clive started from scratch, as did many others with people who were to help our college. I had read a great deal about his work and one day I simply called him up and asked if Sally and I could go up to see him at Cornell University. We fell in love instantly with the McCays and Sally and I were simply goggle-eyed at the extent of their experiments and what they were doing with older people to help them overcome the problems of advancing age. Clive them came to see us in Rutherford and to lead a discussion with our students. Through the years I was to follow this pattern often. I usually found that the man at the head of the parade was the most humble and most willing to help. And in us they found willing disciples eager to work out the mechanics of the instructional process so that our students would benefit. But even more important, the richness of our friendship was something that lent a special glow to our lives.

We bought organically grown wheat through Paul Keene of Walnut Acres Farm in Pennsylvania and the local baker made the bread according to our formula. It was a wonderful bread and every visitor to the College got a loaf with our compliments. Soon the reputation of the bread attracted the attention of a friend of mine, Bill Sherman, who was Vice-President of Hathaway's Bakeries. He proposed that we allow his company to use our formula to be marketed under the name "College Bread" and in return the company would pay the College one cent royalty for every loaf sold, the money to be added to our scholarship fund.

Hathaway's Bakeries was a Boston-based company and like all bread companies began to experience financial pressures because of the lessening consumption of bread by the American public. Soon it had to discontinue its New York metropolitan operations and College Bread was no longer on the market. It was at this time that I observed that there was a tremendous wheat surplus in the United States, and that much of it was being stored in rat-infested holds of old Liberty ships on the Hudson. I wrote a letter to the editors of fifty dailies in the United States stating that people should eat more, not less, bread, but that the bread should be genuine whole-wheat bread. If every person in the United States ate one more slice of such bread a day, there would be no wheat surplus, bread companies would prosper, and people would be better fed.

One device we tried in our Freshman orientation was to have each student keep a seven-day record of everything he ingested, translating each item into vitamins, proteins, minerals, and calories. Totals were divided by seven and compared with minimum daily requirements. This clinical record would be informative enough to any intelligent student, but we further encouraged the presentation of such studies to family physicians for their counsel and guidance.

I had a group of fifty millionaires in New Jersey keep the same record and then followed up with similar studies of fifty faculty members and fifty college office workers. Most of these were the heads of some of the largest corporations in New Jersey and included such persons as Roy Hurley, president of the Curtiss-Wright Corporation, and Leonard Dreyfuss, president of the United Advertising Company. Their secretaries would ask them each morning what they had eaten and drunk on the preceding day. In all cases but one, the secretaries became so interested in the study that they made the same record for themselves. I remember one saying, "Gee, my boss doesn't eat as well as I do!" But the results were substantially the same. All were deficient in calcium, in iron, and in certain B vitamins. The faculty members were slightly better. Should not college students as part of their overall science training be encouraged to adopt a more objective point of view on health and not allow Madison Avenue hucksters to make the decision for them?

At this time began our interest in organic food. Today the incidence of pesticides in our food has finally reached alarming proportions and is drawing the attention of the Federal government. At that time, people thought we were somewhat daffy to be concerned with a little pesticide in our food. At any rate, we tried to buy as much organic food as possible for our cafeteria, even to the extent of advertising for sources of such foods. I also prevailed upon the trustees to purchase a 250-acre farm in Chester, New Jersey. It was my idea to have vegetables grown organically for the students' dining halls. The estate was virgin land and no injurious chemicals had been used on it. We had two cover crops grown and plowed under to enrich the earth naturally, and by that time it must have been one of the richest soils in New Jersey. We set up a compost heap and arranged the first small tract, bringing in thousands of June bugs and earthworms. My friend Dr. Ehrenfreid Pfeiffer, an expert on organic farming, came enthusiastically to advise us. Alas, how naïve I was. It soon became apparent that there aren't any farmers left to work on such a project. We tried custom farming, but even these agricultural contractors were shifting over to housing developers. After a few years we sold the property at a handsome profit for the university and relied on the usual source of pesticide-laden vegetables.

A friend of mine, Cooper Marsh, who had been deathly allergic to pesticides, was wealthy enough to set up his own farm where he bred steers fed on organically grown pastures. He did not fatten them unnaturally and the steers lived a free, natural life the way God intended them to live. We have learned to take fattened animals for granted. Cows are fattened so much that farmers have to hold up their teats with bras. Those beautiful steaks streaked with fat are more tender, but we are paying a high price for this: we are loading up on cholesterol. The great danger is that the fat absorbs pesticides and the more fat we eat the more pesticides we ingest. At any rate, I received the cooperation of Cooper and bought his beef at what was a less-than-cost price but still higher than commercial meat. There was a time when the Fairleigh Dickinson students enjoyed the best beef that money could buy. Did they appreciate it? No. A hamburger is a hamburger, and when you've drowned it with ketchup who can taste the difference?

We even had Walker-Gordon registered raw milk. Here the reaction was different. The students drank so much of this milk that it became financially impossible to continue to supply it. I must explain a little about milk as it comes from the cow. Just as the milk from a mother's breast, it has the best nutritional value in its natural state. If we were to pasteurize mother's milk, it would lose a great deal in the process. The same thing holds true with cow's milk. But here we equate the loss with possible infection, especially undulant fever. But if the cows are carefully controlled and tested every day, the natural raw milk has no chance of being infected and is the best and most nutritious milk product one can obtain. This then was the milk that Fairleigh Dickinson students were getting and which they were drinking in copious amounts. It was ironic that the one item of food that did prove popular, we were unable to continue because of excessive costs.

We in America are undergoing basic changes in our eating habits. Some are good; some are bad. The taste for natural foods is disappearing. For instance, young people do not differentiate between different types of fish. In the fried fish sticks, they taste the fried butter and the ketchup. The same goes for French fries. Most store breads are entirely tasteless, of strawlike texture and almost valueless except for the few artificially added vitamins. Luncheon meats seem to have a high pesticide count—probably because of the high absorption of pesticides in fat. Now because of lower costs but mostly because of weight-reducing fads, artificial sweetening agents are replacing sugar. Sugar was bad enough, but the cumulative effects of cyclamates, we now know, are much worse. In our desire to keep foods from spoiling, we are using more and more chemical agents. What we were trying to do for our students was to teach them to use natural foods as much as possible, to appreciate natural flavors as opposed to artificial flavors, to know the values of good, simple, unadulterated food.

We set up a required nutrition course for all students. Four of us from the faculty and administration attended the full sessions for the first semester. The instructor knew his facts thoroughly and we were anxious to learn from him. We also wanted to observe how the students were reacting. Curiously, we did the same thing in our school of dentistry, but there we had an additional reason. We wanted these young men as future dentists to have a concept of the importance of nutrition in their patients' health. Many students now are on a holy crusade to get more meaningful food and to go back to nature, all of which is good. We accomplished something, but such courses are usually taught on too technical a level for the undergraduate students.

We conducted a few experiments too. We took a group of twenty girls, all of whom had had poor attendance because of illness which included headache, stomachache, overweight, underweight, or general debility. With the help of Dr. Roy Black, who was the consulting college physician, we put them on a sensible diet: good breakfast, low calorie, high protein lunches and dinners, only one cup of coffee a day, cake once a week, no candy. After one month, the effects were remarkable: physical illness disappeared, attendance was almost perfect, the fat ones lost weight, the thin ones gained.

Then we conducted another experiment with two small dormitory units. One was left to have meals as they wished. In the other, we prevailed upon them to try a three-month experiment of having a good breakfast every morning, which, incidentally, they had already paid for, as they had for all their meals as a regular part of boarding fees. Again the effects were dramatic. In the other dormitory, there was the usual run of colds, with most of the students having to stay out two weeks or so, and the usual languorous mid-morning feelings. In the good-breakfast dormitory, there was only one minor case of sniffles that lasted a day and no weight gains during the period.

Some people in the academic field would look down upon such efforts to help young people. I don't know what is more important than health in the lives of young people in our charge. And there is no question in my mind that mental performance and academic progress are to a great extent conditioned by good eating habits.

One important tangible result of all our efforts for good nutrition was the establishment of the Health Research Institute under the direction of Dr. William Smith on our Madison campus. This gave us an opportunity to run nutritional experiments with live animals to see the effects of certain foods on these animals. As I took students or visitors through the various rooms, it was upsetting to see how certain mice, for instance, had mangy hair and other defects if fed on ordinary store bread, and how well they looked when fed on whole grains. And it was even more upsetting to see the production of cancerous tumors where animals were fed on some respectable items of food. But to get the real thrust of the research, one must find out whether the findings are having any effect on the researchers themselves. The last time I asked the question of one of the research leaders, Dr. Yacowitz, he answered that he tried to eat fish three times a week and white meats at other times. In other words, he was eschewing red meats because of the better results with white flesh.

Just to point out how fast this earth of ours is being polluted, just a few years ago we were assuming that fish was the one item of food we could rely upon, especially deep sea fish. Now we find to our horror that fish caught hundreds of miles off shore can have dangerous amounts of mercury. My guess is that we have probably passed the point of no return, that our bodies have accumulated more chemical poisons than nature ever intended, and that it will take major changes in our food habits and many decades before we can reverse the process. Now wouldn't it seem to make sense to encourage college students to bear down on this problem and to point the way to their fellow students?

In the summer of 1969, Gloria Swanson called me in Maine and suggested that I drop in to see a group of young people, all of whom had been using drugs, to observe how they were getting along. On our way back to Rutherford, Sally and I did stop in to see the group in Brookline, Massachusetts. We spent two days observing them, eating with them and discussing their philosophy. We liked what we saw. First of all, they were all off drugs. Second, they were all happy people, friendly toward each other and thoroughly considerate. Sally and I were perfectly at home with them. They had gotten rid of the pressures and anxieties of life and were all working at various jobs and earning a living. They lived in three communal houses under the leadership of Dr. Kushi, a Japanese philosopher and disciple of Dr. George Ohsawa.

They lived on a macrobiotic diet and seemed to be thriving on it. Not one had seen a doctor in two years. They cooked their own food cooperatively, had a far finer palate for good food than most people, and enjoyed their meals thoroughly. What more could you want out of life?

Macrobiotic cooking is basically oriental cookery. Dr. Ohsawa coined the term, meaning "great vitality." The emphasis is on whole grains. It is essentially vegetarian. Fish and fowl are secondary foods. Vegetables and fruit, all organically grown, are important. No alcohol is allowed. They made their own bread and it was the best we ever tasted.

With their enthusiastic assent, I wrote to their parents telling them of our reaction and reported to them our experiences. I told them that it wasn't a diet I would care to follow, but then, why shouldn't they have the right to eat as they wish, especially since they seemed to be thriving on it.

Back in New York, I invited members of the Board of Fellows to sample a macrobiotic dinner. Gloria came over with four of the Brookline community, including Dr. Kushi. We did break the rules and had cocktails, but otherwise we stuck to the principles of macrobiotic cookery. It was a most interesting evening, especially the philosophic discussion that followed. I asked Charles Rodman, president of Grand Union, for his reactions.

"There are so many changes in the food industry, I want to know about food habits of all peoples," he said. "I enjoyed the food and I learned a great deal tonight."

One other business executive, who shall be nameless, demurred. "Give me a great big steak any time," he said, as he sipped his fourth highball.

In all of this discussion one may gather that I was one of those ascetic creatures subsisting on birdfood and raw salad. On the contrary, I was always interested in gourmet food. At the end of prohibition, through the column of C. Selmer Fougner, wine editor of the old *Sun* in New York, we formed the Wayfarers' Club, composed of professional and business men interested in good food and wines. Our first meeting was in the old Hotel Lafayette, owned by Raymond Orteig. Mr. Orteig was the man who put up the $25,000 prize for the first solo flight over the Atlantic, won by Charles Lindbergh. Mr. Orteig and his maître d'hotel, Henri, showered us with every attention. As it happened, the worst snowstorm of the year took place that night. Half of the group was marooned in the small hotel which has since been torn down. But this was purely a men's club. Later Dr. George Hyslop formed a small group of four couples interested in gourmet cooking. We rotated the get-togethers and each couple in turn was the host. By that time the war had broken out. It was our turn, and meat for eight people was difficult to get. We decided to have bouillabaisse. Sally ordered the various kinds of fish plus the lobsters. We were going to simulate the fish stew we had had in Marseilles. It was an hour before the guests arrived. The fish was all over the kitchen. Suddenly, Sally burst into tears. "I've got all this fish and I don't know how to put it together." She struggled through and then put on her evening dress as the guests arrived. In a whisper I told them what had happened. I said, "When I pull out the handkerchief from my breast pocket, start praising the bouillabaisse, no matter what you think." Well, it didn't need any signal from me. As usual, Sally came through with flying colors and the guests were lavish in their praises. As usual in such groups, the men get the ideas and the women have to do the work. The only working male in our group was Manley O'Kelley, a Southerner and an admiralty lawyer. Manley was a sight to behold. He always sported a monocle and as he worked away in the kitchen in his tuxedo he would keep up a running conversation laced with good stories, all with a charming mixture of Southern and English accent.

Later I joined the Ecoffier Society, presided over by Joseph Donon. Still later, when we acquired the Madison campus, which had been the estate of Mr. and Mrs. Hamilton McKeon Twombly, known as Florham, I learned that Mr. Donon had been maître d'hotel and major domo for them. I wanted to learn all about the mechanics of the place and so I called Mr. Donon, who now lives in a comfortable cottage in Newport. Right after the First World War, Mrs. Twombly —whose New York home was at 71st Street and Fifth Avenue, across the street from the great Frick mansion which is now a museum— asked Mrs. Frick whom she could get as a chef for her estate in Madison. Mrs. Frick suggested a young man who had been severely wounded in the First World War and who was looking for part-time work. Mr. Donon went to work at Florham, which was an in-between estate for the Twombleys. They spent the winter in their mansion in New York, their summers in a mansion in Newport, now Regis College. During the spring and in the fall, they would live in Florham, which became the nucleus for our Madison campus.

One of the student clubs that was formed at the college was a gourmet club. The students would seek out restaurants that served outstanding food at popular prices. The boys and girls in this club had a marvelous time. Sally and I used to join them at times. Clementine Paddleford, the food editor of the late *New York Herald-Tribune,* became interested and wrote of their activities.

Good and healthful eating habits need not be dull. They can be very exciting. I still think that a good place to acquire such habits is college. After all, mens sana in corpore sano.

The caller had been referred to me by one of the well-known ministers of the county. He himself was equally well known and his forebears were among the early families of the area. He was a contractor and he wanted to bid on our first new building. I told him that I would be glad to add his name to the list of those presented to our architect for possible bidding. As he was leaving, he said hesitatingly, "By the way, if I get the contract, I'll be glad to take care of you personally." To say that I was shocked would be putting it mildly. Having a new building of this nature was a new experience to me. I assumed that builders of repute would estimate on the architect's specifications as frugally as they could and that especially in the case of a nonprofit institution they would go out of the way to shave costs. It was inconceivable to me that such a devious way of doing business with schools existed in the field. I said something to the effect that the only thing he could do would be to make his bid as low as possible and that the architect would recommend to the trustees on the basis of sealed bids opened at a specified time, which is how it was done. However, the contractor in question was not asked to bid on that building or on any future buildings.

But the experience taught me a lesson and as I mulled over the matter in years to come, I decided that there could not possibly be any collusion of any kind and that everything would be done in a goldfish bowl. The next building was a gymnasium. We had naturally developed the plans with the faculty, which at that time was a relatively small one. We also brought in a committee of students to give us their point of view and to offer us suggestions. We went through several stages of preliminary rough drawings and finally the board of trustees approved the final draft. They approved the list of contractors who were to bid. But at this point, I decided to make the project one of maximum participation by students. A copy of the plans and of the specifications was left on a library table so that the students could have the experience of seeing what they looked like. After all, as citizens in their own communities this would be a worthwhile experience. For those who may not be familiar with bidding procedures, the contractors are usually given a certain number of months, depending on the complexity of the building, to prepare bids. Usually they have to get subcontractors to prepare bids on electrical, plumbing, heating, and perhaps other aspects of the building. They reserve their own general contracting to themselves and at the last minute lump all of these things together to come up with the total bid. Sometimes the same subcontractor will bid for a number of the general contractors who are bidding on the job, in which case that particular cost will be common to the various contracting firms. What should be a very precise and very efficient process is often a very imprecise and sloppy procedure. If the contractor is hungry for work he may stick out his neck and bid low, hoping to make a fast buck. On the other hand, if he is extremely busy, he may hand in a carelessly figured bid, not caring too much whether he gets the job or not. We have found out that in some cases, the wife of the contractor figures out the bid much as she would a home budget. Naturally, as the job gets larger and the complexity increases, there is less occasion for hand-to-mouth manipulating, but considerably more padding in order to offset any miscalculations.

At any rate, I made the opening of the bids to coincide with our college community conference, which was the weekly meeting of the entire college, faculty, and students. Contractors usually want to be present at openings. We placed them on the stage. The architect was, of course, present there too. The sealed bids were on a table on the center of the stage. To the side was a blackboard on which had already been written in columnar form the names of the firms bidding. The students ran everything. One student opened the bid. Another read out the actual bid. A third wrote the amount to the right of the name of the firm. After a while the students would get the hang of the thing. On a low bid they would applaud; on a high one they would boo. Then began the real educational part of the procedure. Students could ask any questions they wanted to. The contractors or the architect would answer. "Why was there a fifty percent difference in bids?" Some contractors would lamely try to answer that one. Curiously, the students were more concerned that the lowest bidders might not be able to deliver. The architect would then explain that a performance bond was required of contractors which would guarantee job completion.

Usually the bids don't come in until a few minutes before opening. The students usually wanted to know why they came rushing in at the last minute. Some bids got under the wire by a few seconds. As a matter of fact, in a later case at our Teaneck campus, the contractor came in after the opening was started. This was a job that involved government participation. At this point, I had to step in. I asked the government representative for a ruling. He had to call the regional office in Philadelphia. While he was away phoning, I explained to the students the reason for clearing the legality of the acceptance of the bid. The representative came back with an affirmative rule—the bid could be accepted if the other contractors agreed. At this point, I asked the other contractors whether they objected to the receiving of the late bid. All but one, as a matter of professional courtesy, agreed to the acceptance of the bid. Incidentally, the reason the particular contractor was late was that he had been to the Rutherford campus. I never saw a more flustered and more puffing contractor. I really felt sorry for him; so did most of the people. But one contractor demurred. He stated in no unequivocal terms that should the late bid be the lowest, he would petition to have it declared null and void. It was a pretty kettle of fish, but a dramatic experience for the students. Luckily, neither the late contractor nor the demurring one had the low bid, so that a legal delay was avoided. But on the question of last-minute bids, the contractors were quite frank. By presenting the bid at the last moment there is no opportunity for any inside peeking. The students were learning the verities of life. The contractors also admitted that, like everything else in life, there is a tendency to leave everything till the last moment. And sometimes the subcontractors hold them up until the very day of the bidding.

On government-financed buildings, the law specifies that the lowest bid must be accepted. On other jobs, the institution has the right to decide which bid to accept, even though it may not be the lowest. We would always go through the ritual of asking the students to vote on which bid to accept. Usually they voted for the lowest. But in a library bid, there was a difference of about $10,000. The lowest contractor stated that he would need twice as long to complete the job as the next lowest bid. The students voted to give the job to the next higher bidder on the basis that it was worth it to them to have the new library finished a semester sooner. The recommendation of the students went to the trustees, who adopted it.

It was explained to the students how the different subcontractors would integrate their activities and, as the building arose, students would often refer to the plans to note some particular item that struck their fancy. At a certain point, usually half through the proceedings, we would take the student leaders through to show them what progress had been made.

On one small building we ran into a serious problem. We had been given a certain amount of money from a foundation to enable us to proceed on a small classroom building. By the time the bids were in, building costs had been skyrocketing so fast that the money on hand could barely meet the cost of the shell according to the lowest contractor, who was very cooperative and really wanted to help us in any way possible. We quickly mimeographed a sheet, which we asked the students to study with their parents, giving three alternatives: first, give back the money and not build; second, put up the shell, then board it up until we had money to go ahead; third, put up the shell and finish the building with faculty and student help. The vote was eleven to one in favor of the third alternative.

Now I had a tiger by the tail. I decided to try my luck with the unions. I went to the hiring hall for the building trades and approached various union leaders. I said, in effect, that since the industrialists were meeting part of the cost, as a matter of public service they should volunteer to donate their services in finishing the building. I offered to establish scholarships for sons and daughters of the union members in return. Five of seven unions agreed to participate in this plan; two were adamant, and without their help the whole plan collapsed. I decided to go ahead anyway. Then, toward the completion of the shell, which included floors and the stairs, I began to bring in students and faculty. In short order, I had a union delegation come in to see me. They were ready to throw the book at me. No institutional building could go up in Bergen County without union help!

I'll never forget the leader of the group. He kept his black derby perched on the back of his head and he chomped at his cigar throughout the discussion. By this time, I was beginning to enjoy the proceedings. After all, I didn't have much to lose. The building was half up and a half-completed building is bound to be finished, one way or another. I explained our finances to them, told them how the students had voted, and pointed out that this building would serve their children. I asked for time to consider.

In the meantime, I consulted sub rosa with an outstanding union leader, who later was to be an advisor to the President. He pronounced my plan unheard of in a tight union county such as Bergen. On the other hand, if there was to be a fight, public sentiment would be on my side. If I had the will to fight, even if I lost, it would be wonderful public relations for the college. I consulted with student representatives and they felt that we should go ahead. I asked for a meeting with the union delegation and I said in effect that, much as I wished I had the money to complete the building with union help, I had no choice but to go ahead on our own. "There will be a fight," the cigar chomper yelled. Well, if there's a fight, I thought, faculty, students, and I are ready to fight. The worst that can happen is that we won't be able to finish the building, in which case we won't be any worse off. Then I said, "I appeal to you to consider volunteering to finish the building as your gift to education. Some of your unions are willing to do it." The delegation stomped out. The next day we resumed our work.

Some of the photographs I prize the most are of students, faculty, and myself in overalls finishing that building. Loyd Haberly, later our dean for liberal arts, was especially good at masonry work; he had lived many years in England and had built his own house. We also found among the students all kinds of expert workers. In some cases, the students did ninety percent of the time-taking and very hard manual labor, such as pulling wires through pipes. At the very end, a licensed electrican would connect the fuse box and certify it legally. Another student was especially good at tiling work. Roland Wank, our architect, looked on helplessly. After all, his six percent commission was based on money expended. But he soon got into the spirit of the thing and prevented many a gaffe that amateurs are bound to make. The heating and plumbing had to be done by local people, but it was a lot cheaper. With the painting, we had a ball and completed everything in record time. The Hesslein building was sorely needed. It gave us eleven classrooms and we all were elated when we got through. It still does yeoman duty; as a matter of fact, it's the first building you pass as you enter our Rutherford campus. But I'm glad I never had to construct a building that way again.

For the students, it was a wonderful experience in the cost and processes involved in public buildings, the role of unions in our economy, and the effectiveness of cooperative action. Many, probably most, of the students had parents who were members of unions. There was no anti-union feeling. It was simply a matter of dollars and cents and of either doing it our own way or not at all. I also showed them the finances of the college so that they could have some concept of institutional budgets. As a matter of fact, I always made this a part of my talks at freshman orientation. College financing is basically simple. Receipts come from tuition, contributions, income from endowment, and lately (although not then), government or state grants of one kind or another. Money can be disposed of in four ways: faculty salaries, nonteaching expenses, capital building, or adding to endowment. One of my colleagues jokingly pointed out a fifth way: simply wasting it. I have always expained to the students and to the faculty that a good formula is that fifty percent of all receipts should go for faculty salaries, twenty-five percent for nonteaching expenses, and the remainder for capital and building endowment. This formula can be varied five percent or so on any of the major divisions, but continued disregard of it will lead to institutional difficulties. Endowment itself has a very imprecise meaning. For practically all of the newer private colleges formed during the last fifty years or so, it simply means building up a backlog of stocks, bonds, and real estate that could be sold in times of stress. Now, as I am writing this, the Carnegie report on the impending bankruptcy of private colleges has just been released. The private institutions have been solving their problems by the simple expedient of raising tuition.

Some administrators and some teachers don't realize the consequence of raising tuition by, let us say, $300, although in some cases it has been even more. For perhaps ten percent of the students, it may not make much difference. If papa is in the $50,000 bracket, a few hundred more can easily be absorbed, although, curiously, even rich parents object. I am convinced that most private universities have priced themselves out of normal existence. They make the mistake of looking at what Harvard, Yale, or Princeton may do. The social prestige of attending such institutions is so great that parents will take out an extra mortgage in order to keep their sons (and now daughters) there. They forget, however, that these prestige institutions have colossal endowments—colossal. The return on a half billion in endowment is apt to be $25,000,000 at least. Divide that by, let us say, three thousand students. The return is $8333 per student before any tuition, grants, or yearly solicitations are received.

Now, these institutions give out enormous amounts in student scholarships, grants, or loans. But in many cases students bring in their own scholarships from the state or federal government, or from private sources. The great bulk of the private colleges don't have this situation. Three things happen when tuition at most private colleges gets too high: first, the student goes to a public institution where the tuition is nothing or very little; second, he may increase his work load to the point where he defeats the very purpose of going to college; third, he may drop out. Now the limit has been reached but no one wants to understand this simple fact. Everybody wants to gobble up 150 percent of the pie.

But I was to hear from Mr. Derby-hat again. Some time later, the old Kingsland House in Rutherford was going to be torn down to make room for an apartment house. Rutherford has a historic past, not as Rutherford, but as a farm area that goes back to colonial times. The Kingsland House probably goes back to 1670, when it must have been a single-room cabin with an earthen floor covered with sand, and a score or so of stones in one end to serve as an open fireplace with a hood and flue to take the smoke out through the roof. It was on an old Indian trail which today is known as Union Avenue. Through the years it has been added to, and the latest transformation made it into a jewel of a Victorian house much publicized in architectural and popular publications. Many Rutherfordians, including myself, hated to see it go. I proposed to the trustees that we buy the house, have it restored, and use it as a classroom, because at that time we were short of space. The trustees agreed and we bought the house and two acres for about $26,000—really a bargain. Just about the same time, another historic house was slated for demolition, the famous Outwater homestead in East Rutherford, one of the best examples of Dutch architecture in Bergen County. In this case, Becton, Dickinson and Company, owner of the house, let us have it without cost, provided that we had it moved. I got one of the best architects in America, Edgar I. Williams, to help me on the problem. He had designed our Becton Hall, he lived in Rutherford, and he was interested in historical restorations. We took down the Outwater house stone by stone, marking each as we did so. We reerected the house next to the Kingsland House and connected the two with an unobtrusive passage-way. We could not afford to keep up the houses as a museum because of the expense, hence the decision to make the rooms into classrooms. We decided to keep as many things as possible in the historical tradi-tion. The students' chairs were patterned after an old captain's chair, actually an antique in our house in Maine. The teacher's desk was copied from an old antique design in one of the art history books. It was a low, slant-top desk seen in many pictures of the colonial period. All of this furniture was made by an old firm, now no longer existing, that had specialized in making reproductions for museums of old historical pieces. Even the wastebaskets were woven specially by an old weaver in Bergen County. We had to compromise on the lights and here we made them as unobtrusive as possible. I made a special trip to William and Mary College in Williamsburg to get ideas, but I felt that their lights were too small for effective classroom use. Then friends gave us genuine antiques which we put into the building: an old bookkeeper's desk, a large wooden cranberry rake, a beautiful grand-father's clock, fireplace andirons, a reverse-painting mantelpiece clock. We tried to develop each room with a special motif. One had old New Jersey maps along the walls. Incidentally, the floors were old-fashioned wood strips, the lower part of the walls, colonial-style planking. An-other room had Audubon prints, even though these were of the nine-teenth century. In another, we had prints by old American artists. In others, old posters, regimental pictures, or old New York prints. It was really a labor of love and our head custodian, Tom McEvoy, since deceased, took pride in "colonializing" the house. It was all done on a slim budget and when we got through we had eleven beautiful rooms at an average cost, including the building and the land, of less than $5,000 per room as opposed to $20,000 per room in our other building. Now everybody has so much money to spend, or at least he thinks he has, that costs per classroom may go from $30,000 to $60,000, depending on how much you can get from legislators or donors. When we got through, the Daughters of the American Revo-lution put a commemorative plaque on the Kingsland house because

Washington had lunched there, and the Sons of the American Revolution put one on the Outwater House. Somewhat later, I was invited to a special meeting of the latter organization and presented with their Silver Medal in recognition of my interest in preserving historic buildings. But two-thirds of the way through, Mr. Derby-Man came in to see me again. "We don't want no building in Bergen County without union help," he announced. I pointed out that no single trade activity involved more than a few thousand dollars, which was true. Most of the basic work had been done. He must have realized there really wasn't much involved because he left with the remark "We'll let you get away with it this time but see to it that you don't do it again." I haven't, for the simple reason that our future buildings were all large buildings and contractors involved in such buildings must use union help. In the meantime, Tom McEvoy, when he learned of the visit, threw every available man he could hire into finishing the building before Mr. Derby-Man changed his mind.

On the outside, I wanted to keep everything in harmony. I kept the old well and had it repaired. Thank God there was no well-diggers' union! Then I tried to get some genuine seventeenth-century tomb-stones to add to the atmosphere. I even advertised in the Portland *Press-Herald.* Then one day I was visiting Plymouth. So many tomb-stones all over the place! I went to see the superintendent of the place.

"Do you know where there are some old tombstones?"

"Plenty of them right here," he replied laconically; "these are stones we don't need."

"Could I buy a few?"

"Sorry, I am required to break them up and throw them away."

Thus ended my search for genuine tombstones. In Ellsworth, Maine, I considered the idea of having a few made up to resemble old tombstones but decided it was too much trouble.

A few years later, I took Helen Meyner, whose husband had been Governor of New Jersey and who was writing a column for one of the dailies, to see the historic buildings. Full of pride on our attempt to preserve history, I opened the front door and what should greet me but the most garish vending machine you ever saw. I was able to have it taken away, but now, as I revisit the building, I notice other encroachments of modernity that wipe out the nostalgic past. Modern desks with garish neon lights, plastic wastebaskets, steel parti-tions. Oh, well! What's the use of being too sentimental? It takes dogged perseverance and one must be everlasting on the qui vive to preserve historical things, and also to preserve the beautiful things of the present. Perhaps it's part of getting old, but I deplore the expenditure of millions of dollars then to realize that the students want to live in sloppy togetherness. In one institution, I saw gigantic buildings on a campus that ran into almost nine figures. But when I saw the classrooms, they were small and prison-like, lighted by blinding neon tubes, with everything messy as all get-out with an impossible bulletin board choked with old announcements that no one read, and with dirty walls—in some cases written upon with still dirtier words. Is this what we're going into hock for?

6

How We Became Involved
in the Health Arts

We were in our summer cottage at Yarmouth, Maine. In the middle of the night, I woke up with a sharp pain in my stomach. I was sure it was an attack of appendicitis. My father had had just such an experience in 1924 when we lived in Winfield Junction, Long Island. We had called our family doctor, who lived in New York, and he had advised some hot compresses, and told us not to worry. Two days later, we had sense enough to call in a local doctor who immediately rushed him to the old New York Hospital, at that time on West Thirteenth Street in Manhattan. They operated on him immediately, but it was too late. Peritonitis had set in and he died from a burst appendix. The memory of this tragedy had haunted me through the years and when my pain developed that summer night, I woke Sally up and asked her to drive me to the General Hospital in Portland. We arrived there as the stars began to fade and were rushed into the emergency ward. As a sleepy intern began to ask me questions, I told him what I thought it was and persuaded him to have a blood count made immediately. The results came down from the laboratory a half-hour later and showed a very high blood count. The intern concluded that I was right and put in a call for an emergency appendectomy. Within a short time, Dr. William Holt, one of Portland's outstanding surgeons, arrived. He noticed that I had a Rotarian button in my lapel and immediately we were on a friendly, first-name basis. As I was shedding my clothes, he said,

"While they're getting you ready, I'll run up to the laboratory and check the test." He came back shortly and said, "You'll be glad to know that your blood count is perfectly normal and that you don't need an operation. You probably have had an old-fashioned stomach-ache."

Sure enough, within a short time I felt absolutely normal. I then asked Dr. Holt,

"How did such an error occur? If it hadn't been for your thoroughness I would have had a needless operation."

"Well," he said, "it is hard to get competent help. The girl in the laboratory is a recent high school graduate and not very knowledgeable." He explained that the blood corpuscles in a square area are counted and then multiplied by a certain factor. If the initial count or the multiplication is wrong, a wrong count results. This is exactly what had happened.

On my way back to Yarmouth, I began thinking. The laboratory is one of the most vital centers of the modern hospital. If the wrong clinical results emanate from it, everything else that happens to the patient may be meaningless; even death could possibly result. I wondered how the laboratories of our New Jersey hospitals compared, and had Sally make a study of seventeen hospitals in northern New Jersey. Through the years, while I have received a great deal of credit for whatever things we accomplished at Fairleigh Dickinson University, it was always Sally who did half of the work, or, as in this case, most of the detailed work. When the results were in, we couldn't believe our eyes. In some hospitals, the laboratory was given to a doctor on a concession basis; the doctor could keep the fees and pay for the laboratory technicians. Training of the technicians was on a very haphazard basis, limited to the odd times when the doctor or head of the laboratory had some free time. Many hospitals were valiantly trying to keep up standards but were hampered by lack of funds in paying for a full-time clinical pathologist. We secured their complete and enthusiastic collaboration. Then we had a meeting of hospital administrators. Most were very happy to be involved in an upward movement for professional advancement—but not all. A few were afraid that costs would shoot up. Some laboratory heads were not ready to lose what really amounted to laboratory slaves, since they were getting practically no pay for supposedly learning the profession. It wasn't an easy transition we were asking for—to step from a nineteenth-century concept of laboratory analysis to a twentieth-century one. We proposed that all the hospitals adhere to the standards set by the American Society of Clinical Pathologists. We then set up a curriculum in medical technology that provided a two-year pre-clinical course followed by one year's work in a hospital laboratory. But there would be a strict course of study during the year, with definite hours of instruction to be guaranteed. The student could then follow with another year at the college or work and get needed credits on a part-time basis. Gradually the resistance began to disappear and I am glad to say that northern New Jersey enjoys some of the outstanding laboratories in the country and that the true importance of the laboratory is nowhere recognized as much as it is in our area.

A few years later I was at a nearby hospital for a kidney stone operation. Since I had been in great pain, my doctor had prescribed a deadening pill. In a half stupor, I lay in bed waiting for the operation. A nurse came with a tray of food and, not quite knowing what I was doing, I began to eat. By pure chance, Sally came into the room, shrieked in horror that I was eating before the operation, whipped away the tray and had it taken away. A little later I was wheeled into the surgery room and the little stone that was causing all the trouble was duly removed and I was returned to my room. For those who are not familiar with this type of operation, let me explain that a post-operational gimmick is to insert a tube into the kidney so that it may drain directly into an outside receptacle. A nursing student came in and attached the tube. Some time later, I looked at the tube but I could see no end inserted into my body. I rang for a doctor who checked the tube and was horrified at the mistake. It could have lead to a fatal result. So, twice in one short operation in a splendid hospital, through carelessness and lack of experience I could have been killed. As I recuperated, I had time to think and I decided to set up a school of nursing in our college. I sent for books on the subject and secured the curricula of established collegiate schools of nursing. I drove the personnel of the hospital almost crazy asking so may questions about the duties of nursing. Let me explain that hospital schools of nursing usually offer a three-year course of study. It is fundamentally a process of learning on the job, with certain classes held during the week. Tuition, if anything, is very low. The trouble with this arrangement is that every time there is an emergency, classes are apt to be called off, and during periods of labor shortage the training program is shunted aside. The result is that there are often gaps in the educational standards. The two errors in my case are good examples.

Some colleges have a four-year course of nursing and eventually we were to have one alongside of the two-year. However, in surveying the graduating classes of the high schools from which we drew students, we found that there was little likelihood of getting any more girls into nursing through this route. But we did ascertain that an appreciable number who wanted to go to college and study nursing too would be attracted to a two-year course.

I decided that it would make more sense to have a two-year nursing course that would combine three things: a general cultural background, pre-clinical studies, and vigorous but directed training and experience in the hospitals. The latter phase was carefully mapped out and worked out in much the same way as the medical clinical experience. For instance, our own instructor would go to the hospital the day before and make a survey of the clinical material available. If she was trying to teach them how to insert a tube into the kidney for post-operational drainage, she would make sure that there was such a patient available. Five or six girls would be brought into the particular ward and given an exact demonstration. Each girl would be given an opportunity to go through the technique with the same patient or other patients. Nothing was left to chance. I encouraged the bringing-in of guest lecturers who were experts in their field. For instance, Dr. Oscar Schwidetzky, who had invented and developed many different surgical devices, including needles, would demonstrate how to use them with minimum pain to the patient. He even had put together a chamois-covered pad that simulated human flesh so that the girls could practice on it before tackling human flesh. When the problem of staphylococcus infection began to plague the hospitals, I brought in Dr. Alfred Globus, who had done a great deal of research on the matter.

The girls would have their clinical experiences in various hospitals, much as the medical students. The pediatrics stretch would be in a hospital strong in the field; the psychiatric experience would be in a mental institution. There was one disadvantage to our set-up. Not being attached to any one hospital, our girls would not at first know the mechanics of the particular hospital they would finally go to work for, whereas the graduate of the hospital school of nursing, after three years, would know the hospital from top to bottom, even if there were gaps in her professional education. But after a few months, the college girl soon caught up on the routine of the hospital.

It really doesn't make sense for a hospital to run its own school—even though there have been many that have done an outstanding job—any more than for a hospital to run a medical school. In many cases it is running a small college within the confines of the hospital. Some hospitals meet the problem by sending their girls to a nearby college for cultural and pre-clinical courses. We arranged a number of such cooperative plans. But a plan still leaves the girls without the benefit of a full-time college experience. She should be able to go to college like any other college major and combine that experience with professional training. The hospital nursing student is cloistered, except for a few hours a week within the college.

Incidentally, getting approval of the State Board of Nursing for the two-year course wasn't as easy as we had envisaged. They had never heard of such a course and to them it merely meant two years instead of three. We pointed out that most of the three years was really given to routine labor that in many cases amounted to cheap slave labor and that our course was packed with enough repetitive practice to make sure that the technique had been acquired. I remember asking, by way of example, "How many times do you have to empty a bedpan to acquire the technique?" Finally I rounded up fourteen high school principals, who appeared en masse at the State Board to testify to the fact that they could not add any more nursing students either to the hospital schools or to the collegiate four-year courses, but they could add to the community's potential by means of a two-year course, since this would tap a new source. One of the principals had evidently had a few highballs before the meeting. He decided to take the initiative and began to sound off. I saw months of planning go down the drain!

However, after much deliberation, the Board decided that they would try this new experiment if the Nursing Department of Teachers College of Columbia University agreed. I swallowed my pride and drove to Columbia to get their help, which they gave me willingly and generously. Later that week I invited Miss Mildred Montag, the Columbia consultant to have lunch with me at the Restaurant Atop Butler Hall, which was then owned by a friend of mine, Fuller Stoddard. Nearby was a table of old City College professors who knew me and eyed my guest. One of them, Dr. George Nelson, who was

The Board of Directors

announces

The Official Opening

of

Fairleigh Dickinson Junior College

RUTHERFORD, N. J.

SEPTEMBER 16, 1942

BOARD OF EDUCATIONAL DIRECTORS

FLOYD RASHUMAN, Chairman
Principal, Han School, Nutley.
EDMUND BURKE, Secretary
Principal, Han School, Lyndhurst.
MAURICE A. COFFEIN
Principal, High School, Wood-Ridge.
GEORGE L. DIERWECHTER
Principal, High School, East Rutherford.
MARINUS C. GALANTI
Principal, High School, Lodi.
GEORGE A. F. HAY
Principal, High School, Ridgewood.
FRANCIS J. HURLEY
Principal, High School, North Arlington.
OLLO KENNEDY
Principal, High School, Passaic.
HUGH D. KITTLE
Principal, High School, Belleville.
JOHN W. MacDONALD
Principal, High School, Hasbrouck Heights.
GEORGE D. MANKEY
Principal, High School, Kearny.
WILMOT H. MOORE
Principal, High School, Rutherford.
HAROLD A. ODELL
Director, Adult School, Rutherford.
FRANK PAPAROZZI
Vice-Principal, High School, Garfield.
CHARLES L. STEEL
Principal, High School, Teaneck.
O. F. THOMPSON
Vice-Principal, High School, Hackensack.

FACULTY

PETER SAMMARTINO, Ph.D.
President.
PAUL E. BAKER, M. A., D. D., Ph.D.
Co-ordinator.
CONSTANCE BOULARD, A. B.
Art.
JUAN DE LARA, B. S. LL.B.
Spanish.
J. FREDERICK DOERING, M. A.
English.
RAY FARMER, M. A.
Secretarial.
BENJAMIN FINE, Ph.D.
Journalism.
MALCOLM FREUDY, B. S.
Photography.
BERNARD W. GREEN.
Sports.
RUTH HOFMANN, B. A.
Social.
ROBLEY LAWSON, M. A.
Vocal Music.
ABRAM MAYER.
Fashion.
WILLIAM A. MESSLER, Ph.D.
Psychology.
J. A. WEISHAMPEL, B. S., M. E.
Engineering.
RALPH WINGERT, B. A.
Instrumental Music.
SYDNEY ZEBEL, Ph.D.
History.

A UNIQUE EDUCATIONAL PROGRAM designed to prepare young men and women for a successful life career.

EACH STUDENT SELECTS A CAREER IN ADVANCE.
The two-year course includes specific preparation for that career with the addition of a cultural background that will equip the graduate for further self-development.

EACH CAREER COURSE IS SUPERVISED by successful men and women who are leaders in that particular field.

GRADUATES ARE PREPARED TO EARN A LIVING, to make a worthwhile contribution to society, and to lead active and useful lives.

Career Courses

SECRETARIAL
JOURNALISM
ACCOUNTING
POPULAR MUSIC
RADIO BROADCASTING
FASHION
ART
SALES MANAGEMENT
JUNIOR BUSINESS EXECUTIVE

ENGINEERING (two year)
LATIN-AMERICAN BUSINESS
PHOTOGRAPHY
AVIATION
MEDICAL ASSISTANT
LABORATORY ASSISTANT
RESTAURANT MANAGEMENT

DRAFTING AND BLUE PRINT READING
SHOP MANAGEMENT
VICTORY COURSES

A non-sectarian, non-profit institution

DAY and EVENING
COEDUCATIONAL

TUITION: Day—$100 per semester; registration $15.
Evening student fees one-half.

For further information write or telephone
THE REGISTRAR.

RUtherford 2-6100

DEDICATION WPAT
SEPT 12, 4:35 to 5:15

CAREER COURSE ADVISORS

PROFESSOR THOMAS ALEXANDER
Teachers College, Columbia University.
EDWARD BELL.
Manager, Empire Hotel, New York.
MAX W. BILDERSEE.
Supervisor of Radio Education, New York State.
PROFESSOR MARY BRYAN
Teachers College, Columbia University.
MARGARETTA BYERS.
Author and Specialist on Style Counseling.
DR. DOAK S. CAMPBELL.
President, Florida State College for Women.
JOSEPH DAHL.
Editor, Restaurant Manager Magazine.
W. L. DUNCAN.
Executive Secretary, Y. M. C. A., Bergen County.
DR. BENJAMIN FINE.
Educational Editor, New York Times.
DR. HAMDEN L. FORKNER.
Teachers College, Columbia University.
JAMES B. FURAY.
Vice-President, United Press.
THEODORE HECHT.
Adult School, Rutherford.
LAWRENCE HUBBARD.
Crossley Surveys.
FANNY HURST.
Novelist.
RUTH HUTTON.
Author on Fashions and Instructor at Cooper Union, New York.
DR. GEORGE HYSLOP.
New York Neurological Institute.
VICTOR JACOBY.
Executive Secretary, New Jersey State Hotel Association.
PROFESSOR ROBERT JAHRLING.
Institute of Public Service, New York.
ELIZABETH MATTHEWS.
Commercial Artist.
BESSIE MACK.
General Manager, Major Boxes Enterprises.
ABRAM MAYER.
Creator of Chic Patterns.
DR. JOSEPH MORROW.
Medical Director, Bergen Pines Hospital.
WILLIAM ALLEN RICHARDSON.
Editor, Medical Economics.
AMY L. SHUSTER.
Former Librarian, Teachers College, Columbia University.
PROFESSOR MILTON SMITH.
Columbia Theatre Workshop.
PROFESSOR FLORENCE STRATEMEYER.
Teachers College, Columbia University.
ALFREDO VALENTI.
Stage Photographer.
PROFESSOR THURMAN VAN METER.
Columbia University.
PHEBE WARD.
San Francisco Junior College.
EDGAR WILLIAMS.
Columbia University; President New York Architectural League.
EDWARD T. T. WILLIAMS.
Chairman of the Board, Becton, Dickinson & Co.
PROFESSOR HARRY B. WILSON.
Teachers College, Columbia University.

The first public announcement in the newspapers of the opening of Fairleigh Dickinson Junior College.

Rutherford. An early faculty tea in the dining hall of the old castle.

Rutherford Campus. Mr. Maxwell W. Becton is being shown around the new Becton Hall by Dr. Sammartino.

Dr. Slavik, who resigned as Ambassador of Czechoslovakia to the United States when his country was taken over by the Communists in 1948, was one of the early foreign visitors to the college.

Dr. Sammartino working with students to complete the Hesslein Building.

Rutherford. The Outwater House was moved stone by stone from East Rutherford and was reerected in Rutherford. The Sons of the American Revolution placed the plaque seen at the right of the door. The building has five excellent classrooms.

Rutherford. The Kingsland House, 1670. The plaque, placed on the building by the Daughters of the American Revolution in 1956 reads: One of the oldest houses in New Jersey on the way of an old Indian trail. George Washington rested here on his return from Newburgh to Princeton in August 1783.

The Rutherford faculty members who were responsible for building the Teaneck Campus. Dr. Sammartino distributing boxes of roses to them in appreciation for their selfless, dedicated service.

Rutherford Campus. Our century candle, lighted only on important occasions. Alas, it burned away before the century had progressed eight years. Perhaps we were trying to burn our candle at both ends. The Fairleigh Dickinson Knights served as a guard of honor.

Rutherford Campus. Our Tenth Anniversary Convocation December 14, 1951.

On the twenty-fifth anniversary of the founding of the college, Dr. and Mrs. Sammartino, in an antique Packard, lead a faculty contingent up Park Avenue in Rutherford.

Left to right: President Lyndon B. Johnson (then Vice-President), Dr. Sammartino, Governor Robert B. Meyner.

Rutherford Campus. When we acquired the new campus in Teaneck, the students of the former Bergen Junior College were invited to the Rutherford Campus by the students of that campus.

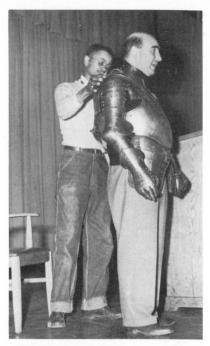

Teaneck Campus. Dr. Sammartino donning a suit of armor to meet the rebellious former students of Bergen Junior College.

The Library in Teaneck with the aluminum relief, William Zorach's masterpiece, which we bought for $20,000 after a Texas Bank had already paid $150,000 for it.

later to become my Director of Libraries, sent a note with the message, "Who's the swell chick with you?" I showed the note to Miss Montag, who laughed and said, "Any time a woman over thirty is referred to as a swell chick, she has reason to be happy."

Now, as I visit friends in various hospitals of northern New Jersey, there are always any number of nurses whom I meet in the corridors who are graduates of Fairleigh Dickinson University and I remember how it all came about because of two mistakes that nearly cost me my life.

7

Strange Ones on the Faculty

One of the first instructors I had at Fairleigh Dickinson was Dr. Eugene O'Neill, Jr., son of the playwright. He taught a course entitled Masterpieces of Literature. He was extraordinarily well informed in all branches of literary creativity, and he knew how they affected our daily lives. Wouldn't it have been foolish for me to tell him what books to teach? I merely told him the obvious: to stimulate his students to read and read, and to make the books he chose come alive to them. He selected the two books a month the students were to read. I disagreed on only one: Gibbon's *Decline and Fall of the Roman Empire,* not because it wasn't an important book but because I felt most of the students would not wade through five volumes of the original edition. Now there is a one-volume abridged edition, but back in 1942 there wasn't. Gene was humble enough to admit that of all the books he had chosen, Gibbon's was the only one they did not read.

Any literary masterpiece is an approach to life and in the hands of a competent instructor can lead the student to discuss and appreciate the insights of philosophy, history, social customs, and the contemporary scene. Over the years I have observed that the outstanding faculty members do this very well. They can do it better when the particular book is their favorite. I know of one faculty member in another institution who had written his doctorate on Bossuet, the French writer known for his funeral orations. He did try to use Bossuet as an integrative force and as an instrument of comparison in everything he taught. He was an introspective and lugubrious individual and none of us was surprised at his choice of Bossuet as a lifetime interest. He seemed to have few friends and, in our happy-go-lucky departmental group, he rarely broke into a guffaw or told a lively story. But perhaps he was right in his choice of funeral orations. It is at the time of death that we cast aside the trivialities of life and pause for a few moments to think of the greater meaning of our existence. I never knew of the effect that his single-minded emphasis on Bossuet had on his students. Perhaps he was able to bring them up to his level of comprehension of the meaning of life and death. I personally think that the subject of Bossuet is too limiting and too confined for the exuberant spirit of most undergraduates. It is bad enough trying to excite them with Racine, Corneille, and even Molière. To me it would be deadening to make Bossuet the central theme. However, this is the sort of exaggeration that occurs because of the specialized research that is required in university circles. On the other hand, I wouldn't object if the integrative author were to be Shakespeare or C. P. Snow or Alfred North Whitehead. As a matter of fact, I remember Whitehead's saying that the courses he gave were Whitehead I, Whitehead II, and Whitehead III.

But back to O'Neill. He was fascinating as a lecturer and as a man. He was a person of violent temper. He had a house in upstate New York where he lived with his wife in what might be described as a tumultuous menage. On one Thanksgiving, in a vociferous shouting match, she left him with the turkey, which he proceeded to eat entirely by himself, more or less in the fashion of Henry VIII. On another occasion Gene and I went through a gargantuan meal at Luchow's on Fourteenth Street in Manhattan that surprised even the German waiter, who was used to heavy eaters. It took me three days to get over that binge. I remember vaguely discussing such disparate books as Fielding's *Tom Jones* and Macchiavelli's *The Prince,* both of which he was teaching at Fairleigh Dickinson, and we spoke as if both authors were living right down the block.

Gene had only one bad habit: he would come to class feeling slightly high. Finally I summoned up enough courage to speak to him about the matter.

"Gene," I said, "couldn't you save your drinking until after your classes?"

"Peter," he replied, "I have to come from New York to Rutherford. How can I possibly get through those terrible meadowlands without a couple of martinis under my belt?"

Through the years I have learned to put up with the idiosyncrasies of faculty members, especially if they were outstanding as scholars and teachers. In the final analysis, a teacher gives the student an approach to life. The student has the right to know the general content of the course but the instructor, if he is of genuine stature and not a phony, will be in a far better position to arrange his own material within the framework of the syllabus published.

There was Dr. Benjamin Cohen, former ambassador from Chile to the United Nations, later Under Secretary at the United Nations. When he retired at the age of fifty-seven, I invited him to join our staff. He taught our Contemporary Society course. After a few months, he came to me and said: "Peter, I shall teach whatever you wish and in any way you wish provided it is in my field of competency. But with my experience all over the world, I can give the students something that no book has yet printed." I told him to disregard the syllabus that a faculty committee had prepared and to take the students through the thrill of history as it was being made. He made the students literally peep in on the contemporary scene as no one else could have done it. He would tell them about motivating factors that were responsible for world decisions. He would give them inside stories of the personalities involved in international matters and how personal piques often led to strange political bedfellows. He had a keen sense of economics and showed how financial stakes often were in back of political decisions. He pointed out in example after example the actual differences between public pronouncements and behind-the-scene maneuvers. One might think that there was cynicism in everything he taught. There wasn't. He explained simply and effectively what the world was like but, more important, he tried to exhort his students toward a higher idealism of service to one's fellow human beings. He made them read books, too, but he also taught them to pore over little-known reports and to compare the printed work of newspapers with the facts that few people knew.

Another unorthodox instructor was Jimmy Roosevelt, the son of the New Deal President. I had him give a special seminar at Fairleigh Dickinson University for a few years. He would prepare his lectures and draw heavily upon his own experiences as assistant to his father. Then he would have a special guest at each lecture, perhaps the then Secretary of Labor, Arthur Goldberg, or perhaps his own mother, or perhaps the then Assistant Secretary of State, Harlan Cleveland. Good speaker that he was, perfectly at home in all sorts of situations, he was literally scared stiff in the classroom. Before he started he begged me, "Peter, could you stay in the room and tip me off if I do anything wrong?" I was only too glad to do so, although it knocked out two evenings a week at a time when I was already working fourteen hours a day. He was a natural-born teacher, even though he was scared. But a few suggestions here and there made him even better. Did he resent my criticism? On the contrary. He had confidence in me and he knew I wanted him to succeed. And he did. He made his students touch hands with history.

Another example was a great Afghan scholar, Dr. Najib Ullah, former Ambassador to Washington and the brother-in-law of the King of Afghanistan. Najib was one of the gentlest souls I have ever met. I observed his teaching. He prepared well and knew his subject, Asian History and Philosophy thoroughly. But he was scared of his students, who were in turn awed by such a personage. The students would all sit in the back of the room so that Najib would have to talk over an abyss of empty chairs. I met with him after the first class and said to him, "Look, you are the leader in the classroom. Have the students sit near you. And talk to them with as much confidence as you did in your diplomatic pourparlers."

"On peut faire ça?" he asked. We alternated between French and English.

"Naturellement!"

He took courage and the students responded sympathetically and willingly. Soon he was one of the most popular professors. All he needed was a few hints to give him courage.

But we had some faculty who didn't work out. One person had a doctorate in business administration but somehow wound up in the field of psychology. Somewhere along the line he had become interested in business psychology, which is really the application of general psychological principles to business. The transition to general psychology was an easy one.

Very often I would try out a person in a single course to see how he worked out. I would drop in on him to see how he was getting along. I followed this procedure with this man and dropped into his class to observe him. I felt that he was a born teacher. He had one illustrative story after another and he had everyone in the room, including myself, in stitches. The man was a natural-born comedian.

As a matter of fact, he had a sizable income as a convention and banquet speaker and commanded honoraria far in excess of any that were offered to most college presidents. He was conscientious, too, about correcting students' papers. The whole trouble was that his humor almost completely blanked out any principles of psychology he was trying to teach. The students were so busy taking down his jokes they had no time to seize the psychology he was supposed to be teaching. In spite of his popularity, it was evident to his chairman that he wouldn't do.

But then something else happened which tipped the scales. He tried to make dates with male students and we soon had repercussions from the parents. Female students would have been flattered, especially since he was a bachelor. But our mores are such that at least in academic circles contact between members of the same sex is taboo. Too bad. He had the makings of a good teacher if other things had been in balance.

When we started, to all intents and purposes, we were a girls' college. As a matter of fact, we started in September of 1942 with 59 girls and one boy. One of the subjects that was to be taught was dietetics. The war was getting into first gear. Just before we opened, the person who was going to teach dietetics resigned to take an important position with the Red Cross in Washington. A day or so before the first class, a lady presented herself and asked whether there was a position teaching dietetics. She said she had her Master's degree and that she had owned one of the best restaurants in Washington. You can imagine how happy we were to acquire her in the nick of time. After a few weeks she said she wanted to set up a cooperative hot lunch for the students. Of course, we gave her carte blanche and were gloating over the prize we had. For thirty cents, students received a four-course lunch. Then expensive kitchen equipment began to arrive at the college. When I asked who was going to pay for it, she said simply that she had made arrangements to have everything on a pay-as-you-go basis and that it would come out of monies received. We had a financial genius on our hands.

Then strange things began to happen. The lunches began to take more and more time. The teaching schedule was disrupted. The meal finally served seemed to bear little relation to what had been announced. One day, I received a phone call asking for routine information about the "restaurant the College was opening in town." We had no such plans. By this time I began to have some doubts. I called in our lady. Up to this point she had been a rather sweet and demure person. Of a sudden, she flared up and, with fire in her eyes, she said, "If you can not have faith in me I shall leave forthwith." Well!

There weren't too many spots in town where a restaurant could be started. I soon found out that she actually was planning on such a move. I called up the local State Senator who had the lease on the building. He upbraided me for not having enough imagination and faith. The lady in the meantime enticed one of our students to go into business with her, and were we short of men students! (By then we had a few more). I called the parents and they were proud of their boy who was on the way to making a lot of money running a restaurant. Then I began to receive bills for equipment. Well, in about two weeks, the project blew up. The lady left town owing bills to half of the citizens in Rutherford. I asked the executive of a large equipment house what prompted him to extend several thousand dollars' credit to an unknown person. I'll never forget his answer:

"Search me. I just didn't think a sweet lady like her would ever try to put over a raw deal."

We sent back all of the equipment sold to the College and went back to sandwiches. In the meantime, our routine check disclosed that the lady had never received any college degrees and that no restaurant such as she described existed in Washington. I learned that lesson quickly enough. Look out for crackpots. Look out for bargains.

8

Sex and the Higher Education

"*But Doctor Sammartino,* the other girls give it away free. I can't afford to," she said to me.

This is what had happened. The owner of one of the private homes in Rutherford had called me. She was a widow and every year she let a spare bedroom in her home to a student. She said that the girl in her house—let's call her Dolores, which was not her name—was receiving late calls from men, and she just didn't like what was obviously going on.

I called in the girl and asked her to tell me the truth. At that time —how far back in history it now seems—it was considered perfectly in order for an administrator to be interested in a student's life while on campus. Dolores was a first-generation girl of Slavic stock. She wasn't particularly pretty nor did she have the high gloss of a sophisticated girl from a well-to-do home. Her father was a factory worker. From her address I gathered she came from a small mill town comfortably far enough from the college. But, as I remember her, she had a fascinating peasant face. I liked her because it was for people like her that we had established the college—young people who had no money to go away to a swank college. A little taller than the average girl, she always seemed to be the type of girl who was ready to do things. She was dressed neatly, not in a flashy manner. She was serious and yet capable of smiling pleasantly. One felt that she had outgrown childhood and that college meant a great deal to her. Because of her height and mature look, I had the impression that she would not appeal to boys her own age. Had she had the services of a special couturiere and a Fifth Avenue beauty salon, she would have been the unusual type that might have been featured in a women's magazine.

The young lady had had an outstanding high school record and had a very creditable college record. In college, most of her courses were in the liberal arts field, with English as a major subject. She also took typing in addition to a regular program because she felt that it was important in getting a job, although eventually she wanted to go on for her master's degree and go into teaching. She had come to Fairleigh Dickinson University on a partial scholarship. She represented the type of student who twenty years ago would never have had a chance to go to college.

After much questioning she finally admitted that she went out with men for money. "But," she went on to explain, "I think too highly of my school work. I could get a job in a restaurant and work my head off for $1.50 an hour six days a week, try to squeeze in my college work and wind up as a mediocre student. But I want time to do my studying properly and I am very serious about assignments. I go out with one man every Friday night. I charge fifty dollars. I am extremely selective and I require the man to have proper contraceptives. My first few experiences in high school were for free to classmates. I soon saw the shabbiness and the ridiculousness of that type of sex. I see my college friends following this path and while they may get husbands through this route, I feel that ultimately I have a better chance. All of my clients are much older and they treat me almost as a daughter. I am the nearest thing to an American geisha girl, and the funny thing is that I kind of like my clients. I perform a service for them and they are helping me through college. Now you may throw me out of college but I will tell you that I intend to keep on doing the same thing until I get my master's degree."

I pleaded that she stop her extracurricular activity, but she was ahead of her generation. It was her life and she was determined to reach her goals in her own way. As to the matter of morals, she pointed out that many of the college girls engaged in sex as she did but they did it for a good time or for gifts.

Of course, I asked her to leave. I am sure that wherever she transferred to she racked up a good record. She was alert and well-informed. I am almost as sure that she eventually got married and made an intelligent and loyal wife.

In the light of the present day Women's Liberation movement, my mind has often drifted back to Dolores. Amateur psychologist that I think I am (and as we all think we are), I believe there were two reasons for her action. The first, obviously, was the urgent need for money. Twenty or thirty dollars more a week beyond what she would have earned at the usual type of part-time job made a great deal of difference to her. I wouldn't be surprised if she contributed part of her earnings to her parents. Two, she was out of the circle of the pleasure-seeking young people around her. At college age, a year's difference in maturity creates an almost unfathomable gulf. All told, she had wrapped up everything in a comfortable package and was enjoying it to boot.

We had another experience with a foreign student from a West African country. In this case, we had arranged for her to room and board with a minister near our Teaneck campus. She had received a tuition scholarship on the urgent recommendation of the Ambassador from her country to the United Nations. When she arrived for registration, we saw immediately that she was different from any we had had before. A little heavier than the average student, she had none of the introverted, shy look of most foreign students who arrive at a new country for the first time. She wore a colorful native dress, was full of bounce, and was ready to break into merry laughter at the slightest provocation. She looked more the type of the incipient stage entertainer ready to belt out the blues.

After a few months we got distressing calls from the minister and even more distressing reports from the guidance office. The minister said she was out almost every night, apparently going to New York. She would come home late, usually in a chauffeured car, or sometimes she didn't come home at all. She apparently had lots of new dresses and began to drip in costume jewelry. She had an expensive hi-fi, and used it all the time when she was home. She also had a typewriter, which she never used. A few more gadgets like a tape recorder, a television set, and a radio cluttered her little room. Whatever electronic device could be bought and brought home, she had it. A check with her instructors brought out the fact that she had attended a few sessions and had dropped out.

In order not to create an international incident, we sent her a few notices to appear for consultation, all of which were disregarded. Then we notified the Embassy that we would have to drop her. Within a week, we received a letter stating that she would be out for an indefinite period due to illness and an impending operation. One night, a big black limousine drew up in front of the minister's humble dwelling. All the dresses and all the electronic gear were taken out by the gay damsel. I met the Ambassador at a few cocktail parties but we never mentioned the incident. But many months later at a small town in New Jersey they were celebrating United Nations week. I was invited to speak as a former president of the New Jersey Association for the United Nations. The Ambassador in question was there, and so was the young lady, who by this time had recuperated from her operation.

Some time later, I was discussing this matter confidentially with an American foreign service officer who knew the particular country well. He said, "Peter, you are too Victorian and you should be more understanding. A young lady in the particular country in Africa does not attach that much importance to sex. Life is to be enjoyed whenever the opportunity offers itself. She was a child of nature and why should you deny a hard-working ambassador a few pleasures while he's serving his country? In his country, a man can have more than one wife and if he arranged to have a girl friend help him to while away the tedium and the stresses of diplomatic life in a foreign country, why blame him? Do you think that some of my colleagues don't have similar arrangements in other countries? If you really wanted to serve your own country, you should have arranged for the girl to stay registered. Now the poor guy has to worry how to have her stay in the United States or he has to have her go back." By the time he was through he almost made me feel that we had done the wrong thing.

"Oh, by the way, if you come across a few hundred acres suitable for a college campus let me know." This was Ed Williams, chairman of our Board of Trustees, speaking to a real estate man. Ed lived in Ridgewood, New Jersey, and was commuting to his new office in Morris Plains. The daily trip was irksome and Ed was asking the realtor to find him a home in Madison. As an afterthought he made the remark that was to result in the birth of a new college. It is out of such chance happenings that sometimes a new institution is born.

The agent did find a few hundred acres at Florham, but the trouble was that they all had buildings on them. I felt that we weren't ready for a third campus and we certainly didn't have the money to change the buildings over to institutional use. At the May 1957 meeting of the Board of Trustees, in deference to my feeling about the matter, it was decided not to consider the purchase of this estate. It was a half-hearted decision and I remember Dick Dickinson's saying, "I think we've made the wrong decision for the right reason."

I had been against the acquisition of this estate. I had proposed to the Trustees that we buy a few hundred acres in New Jersey for possible use in the far future and to buy it before the builders got it all. I remembered the story of Columbia University on Madison Avenue and 49th Street in Manhattan. Things got crowded and they decided to move uptown. They could have had everything from 110th Street north for very little. Instead, they were unsure that anyone would want to go that far uptown and they contented themselves with relatively few acres at Morningside Heights. Many decades later, this cramped campus was to create major tragedies.

Later, I went away to Maine with a great load off my mind. But I hadn't heard the last of it. Hi Blauvelt, one of the trustees, had fallen in love with the place. While I was away he worked on the owners, who had bought it from the Twombly estate, and he got them to lower the price by $500,000. He also got them to spread the payments over a seventeen-year period. This they were glad to do, because taxwise it was to their interest. Ed Williams was not hard to sway; neither were some of the other trustees.

The scene shifts to Maine. Bill and Annette Schieffelin were having dinner with us at East Sullivan, Maine. The phone rang. It was Ed Williams. Usually he was very direct and forceful. This time his voice was hesitant and almost apologetic.

"Peter, if you don't have to worry about the financing part of it and if you can delay using the property for as long as you wish, would you still have objections to acquiring Florham?" he asked.

"Ed," I replied, "what can I say? I sense that the trustees want to acquire this estate."

I returned to the table and mentioned the matter to Bill.

"Florham?" he exclaimed, "Why, Mrs. Twombly was my great-aunt! Annette and I used to spend many happy weekends there." He then proceeded to tell us of the society at the turn of the century that frequented the estate known as Florham, which was used only in the spring and fall because, as I have already related, the Twomblys had a winter home in New York and a summer residence in Newport. Every Friday afternoon, thirty or forty guests would arrive for the weekend. Each couple would usually be assigned a suite and Donon, the major domo, with a staff of chefs, would work all week to prepare for the weekend meals, each one of which would be a Lucullan feast. This then was the property that had been acquired by the College.

At any rate, in mid-August of 1957, the call had spoiled the sweetest part of my vacation in Maine. I had to come down to sweltering Rutherford for a special meeting I had called of seventeen administrators and key faculty men. The flash call probably spoiled things for them, too. They met at my home at eight in the morning and naturally wanted to know what all the mystery was about.

"Just get into your cars and follow me," I replied.

They followed me to Madison Avenue in Madison, New Jersey, long known as the Rose City of America. Finally, after passing St. Elizabeth's College, we came to a beautifully bricked wall, and when we saw two massive pillars we pulled into two open iron gates. After riding through a beautifully kept road and underneath a railroad trestle, a magnificent building came into view, reminiscent of Hampton Court in England. As we reached the impressive courtyard, I stopped. Sally and I got out and the rest followed suit.

"Well," I said, "this place is ours. Now what in hell do we do with it?" There was a low whistle, a gasp of incredulity, and then the questions began to tumble forth.

Slowly, I told them the full story. This was the estate known as Florham, built by Mr. and Mrs. Hamilton McKeowan Twombly. Florence Twombly was the granddaughter of Commodore Cornelius Vanderbilt. Florence and Hamilton Twombly had built the estate, one of the most beautiful in America, and the name Florham came from Florence and Hamilton. They had spent millions on the palace, two million on the grounds alone, and this at a time when gardeners were getting a dollar a day. They had built another edifice for their daughter Ruth. Since she liked tennis, it had the most expensive clay court in America plus a dazzling indoor swimming pool with frescoed walls and surrounded by exotic trees and shrubs.

Most abandoned estates look abandoned. This one did not. Every blade of grass had been carefully cut. The marble halls within the palace were glistening. The massive mahogany doors had been French polished by hand. Most of the bathrooms had luxurious dressing rooms attached. But not a stick of furniture was left. We met all day sitting on the carpeted grand staircase. If we just let it be until we needed it, the place would deteriorate. After all, when they lived there, the Twomblys had 128 employees. We had to do something. We had to spend money, and we thought we might as well spend it for students. That's how our Madison Campus was born.

To this day, if I had to make the decision again, I still would feel the way I did in May of 1957. And yet everyone who has seen our Madison campus marvels at it and comments on what a beautiful estate it is. I was to write a short book on Multiple Campuses and I am sure that most people thought I gloried in so many campuses. I didn't. I felt then, as I do now, that private institutions had best remain small and contain their activities according to the monies they have available to meet deficits.

Sally and I were having a leisurely cocktail when the front door bell rang. We heard the housekeeper answer the bell, open the door, and quickly bang it shut. She came in frightened and said that a group of "bums" was outside. I went to the door and through the side glass panels I saw that it was a group of students with a faculty member, Winifred Bush. I opened the door and we all burst into laughter. I knew immediately what was up, even though our housekeeper, judging by the expression on her face, looked as if she were questioning the sanity of her employer. This was a group of students who were going to live in the flophouses of New York for three days. They each had exactly two dollars to live on. Winifred, who was just about as spunky an instructor as anyone I have met, was going to share the experience with them. All this was long before students began to protest about "social evils" and to demand "relevance."

I had proposed to the faculty, as a matter of basic educational principle, that they encourage their students to learn firsthand about civic and sociological conditions. Among the possibilities I had suggested a visit to the Bowery to study the effects of alcoholism. At that time drugs had not achieved supremacy as an evil. Alcohol was the great demon.

The students, before they set out on their Bowery expedition, only wanted me to see how they looked. Winifred had dressed them in old and ragged clothing and they actually looked like derelicts. They had one of the most vivid and most harrowing experiences of their lives. Some of them literally vomited at the sight of the food they were to eat.

None had ever imagined the utter degradation of a flophouse dormitory. They were profoundly moved by the realization that the dregs of humanity before them were people, just as they themselves were.

Winifred had reported to the local police station and told the authorities what the group was doing. She had a way about her and soon she had the police enthralled. The captain on duty said, "Look, we were due to have a periodic round-up of prostitutes in a few days. Why don't I have one tomorrow so that the students can see what happens in a Night Court?" Late the next night, the students were on hand to see, firsthand, how justice was meted out to the lowly.

In the crush of things that keeps a president's office busy, I forgot about the group until three days later, when I opened the *New York Times,* and there, on the break page, was the story of our students and their visit to the Bowery. The afternoon of the same day, as I was crossing the campus, I saw a group of disreputable looking men who clearly were too old to be our students. I asked them what they wanted. In those far-off days, nonstudents did not get involved in campus affairs. But almost immediately, two of our students who had been to the Bowery ran up to intercede. They explained that they had found conversation with the "bums" on the Bowery so interesting that they wanted them to speak to the other students. This was carrying out another principle I had always felt strongly about: to bring outside speakers from all levels of society, not merely "names," into the classroom. But, frankly, I had never envisaged the speakers as including flophouse gentlemen. The six visitors, our guests, deserved our hospitality. I took them into the cafeteria and I felt that I couldn't go wrong in ordering a good protein dinner—roast beef, to be exact. They ate ravenously. In the meantime the Bowery explorers were alerting other students and soon the circle of onlookers became larger and larger. When our guests had eaten their fill, finishing with apple pie à la mode, the inviter group of students started an impromptu seminar. The Bowery habitues were not at all hesitant to tell how they had fallen on evil days. One of them had a doctorate in history. As they spoke, they seemed more and more, in spite of their clothing, no different from the average decent citizen.

Quite apart from the derelicts whom I have mentioned above, the students ran into a nondrinking professional hobo. They wanted me to meet him. The students had already invited him to speak at the campus, but, unfortunately, it was at a time when I couldn't attend the discussion. I invited him to come to my home and, by chance, one of my trustees was there too, Leonard Dreyfuss, president of the United Advertising Company. The gentleman was dressed in a light tan khaki suit with belted jacket. He was a tall, impressive-looking person, very articulate. He had had printed small booklets explaining his philosophy of life. Leonard and I found him fascinating. He explained that he wanted to be completely unfettered. He always slept in the open air as human beings were intended to live, protected, of course, from major inclement weather. Once in a great while he might compromise on this stance by sleeping in what we usually call a flop house. To admire the wonders of nature was his great pastime. He loved trees, flowers, birds. But he also liked to observe people and to see different localities. He enjoyed walking above all things, whether in the country or in the city. He read deeply, one book at a time, which he would carry with him. It was usually a paperback, and when I saw him he was reading Dostoevski's *Crime and Punishment.* Sometimes he would pick up for a dime or a quarter a book in the used-book stores on Fourth Avenue in New York City, or in some other city. He ate simply, usually fruit and vegetables, usually raw. Out in the country, he might build a small fire, scout-style, and cook a few potatoes or a mélange of green vegetables in water. He always carried with him four things: a little salt, a few matches, toilet paper, and a small bar of soap. Once in a while he would have a bowl of soup in a soup kitchen or in a low-price restaurant. He savored each fruit thoroughly. He was an authority on apples, peaches, pears, and all kinds of fruit, having picked them ripe in practically every state of the Union. He reminded me of another great vegetarian, Percy Grainger, the great pianist. Mr. Grainger would travel all over Europe, wearing a somewhat similar suit and carrying a small knapsack. He would subsist almost entirely on hard biscuits similar to our Ry-Krisp and cheese. "Don't you get tired of such fare?" I asked. "When I get tired of cheese and biscuits, I stop eating for a while. Then I find renewed joy in my simple fare." When he had dinner at our home, Sally was at her wits' end trying to figure out a delectable choice of vegetables and fruit that might do honor to a great composer and pianist.

But to get back to our hobo friend. He worked only to earn as much as he needed to meet his simple wants. Most of the money he earned came from the sale of his little printed tracts. If he had two dollars in his pocket he had far more than he needed to meet his living expenses. In a pinch, he could always take a job for a day or so at some simple task. The world was always full of wonders and surprises. He was always meeting interesting people and having vibrant discussions. He went to all kinds of churches and sometimes slept in them. He loved to go to art museums and he spent endless hours in libraries. He knew practically every important park in the country. Leonard and I were fascinated by his philosophy of life. We asked him if he ever got sick. He replied that he had enjoyed good health for as long as he could remember and hadn't seen a doctor in twenty years. We asked him if he had put any money away. Yes, he had built up a savings account of about $100 and had never had occasion to draw upon it, even when he needed a few dollars to have his little booklets printed. Didn't he miss friends?

"I am never at a loss for human companionship," he replied. "Actually, I have buddies in different cities and we run across each other on our peregrinations. I am never lonely. Sometimes I want to be alone to read and to enjoy the beauty of this world."

"What about women?"

"I speak to women, and once or twice a year I may have a discreet affair with a lady companion. Just now there are three women in different parts of the country to whom I can always turn. But I don't let women tie me down."

"Don't you have any possessions at all?"

"None beyond the little bundle I carry effortlessly. One extra shirt and one extra set of underwear and socks. The bundle serves as my pillow when I sleep. I don't let possessions possess me."

I was to hear the same statement from my friend Perle Mesta a few years later, when we were her guests at the American Embassy in Luxembourg when she was Ambassador. "But Perle," I said, "you don't have to be tied down to houses if you have enough stocks and bonds to give you the freedom of rental." Whereupon, the guests roared and one gentleman immediately pronounced it a fifth freedom—the freedom to rent.

The hobo went on, "Oh, I carry a simple mess kit as any camper would and a small lightweight poncho which I rarely use. I have never seen anything beyond these few necessities that I would like to possess. I feel a joy of living, a natural euphoria, an absolute lack of tension, a feeling of complete trust in the beauty of tomorrow. I enjoy everything I do. I enjoyed speaking to your students and I'm enjoying speaking to you, even though I think you have complicated life for yourself to no rational purpose."

Leonard asked him about old age. Who would take care of him?

"Old age?" the hobo replied. "I'm old now. I'm over seventy. (He looked about fifty-five or so.) I don't need anyone to take care of me. I am self-sufficient. And anyway, do you think anyone wants to take care of people who are physically or mentally old? Some day I will feel the end is near. I always carry a couple of pills. One swallow and the end will come quickly and painlessly. A clean end to an interesting life."

On that macabre note, the conversation palled. He had reduced life to its important essentials and while we weren't ready to buy his idea of working only to pay for basic needs, we did stop to think that perhaps there was no point in overworking ourselves toward an early end and filling our hours with endless worry. This man was not too far removed from the ancient philosophers and even some contemporary ones. Again the students had made it possible for us to learn something.

Through the years, other students repeated the experiment in somewhat different form. In one case, a group rented a small apartment in the slums just to see how the neighborhood functioned. They learned that while a small group of men are ne'er-do-wells and loaf, the great majority prefer to work and earn their living. They began to appreciate the matriarchal system that develops in certain cases. They saw only too well the nutritional deficiencies of children and, I hope, the reasons for my urging students to have some concept of basic nutrition.

Another group went to work as migrant workers. Thereafter, the backbreaking, monotonous work of the migrant had real meaning for the students. They were dismayed by wretched, haphazard living conditions and by the low level of educational services. But they also realized that they as consumers were reaping the benefit indirectly of a system that in order to function economically relied on the subservience of human beings.

These experiences must have led to some understanding, the first step in having compassion for our fellow human beings. The average student living in a neat, comfortable suburban home cannot possibly realize the human misery that exists in the ghetto areas unless he sees for himself. One of our Edward Williams College girls, cognizant of the problem her fellow student, a colored girl, was having studying, would take care of the girl during weekends and during examination periods, simply because the less fortunate girl lived in a small four-room apartment with five other children. It was impossible for her to study or to read.

11

Problems of Foreign Students

"How is my little gel getting along? I got tree dollars for my little gel. You give to her."

This was a mother in Freetown, Sierra Leone, talking. We had arrived in Africa on a quasi-official tour as guests of two West African countries. But we decided to see a few more countries in that part of Africa. In Freetown we looked up this particular parent. The entrance to the house was not much more than a badly chipped block of cement set as a step to a battered door. The interior to the house was dark and gloomy, and the few pieces of furniture inside would not have been displayed for sale in the most decrepit used-furniture shop in America. Near the stoop was another daughter, beautiful in face and in figure, and topless. Evidently she had learned a little about Western prudishness, for as we entered she made a vain attempt to cover her breasts. The house was not much more than a plaster hovel, set back of a regular house on the street and reached by a narrow alleyway that was simply hard-packed earth. Old tin cans and stray pieces of garbage littered the alley. It was depressing. Was this where our foreign students came from? What a tremendous change it must have been for such a student to be plunged into American middle-class suburbia! Were we really helping the student in plunging her into a glittering new world only to plunge her back into these sordid surroundings?

Wherever Sally and I went in various parts of the world, we tried to visit families of foreign students at Fairleigh Dickinson University. We wanted them to know of the personal interest we had in their sons and daughters. But more important, we wanted to learn about the background of the students and something of the problems they had to face back home. All in all, we visited homes in twenty-one countries from Indonesia to Liberia, from the Netherlands to Brazil. Each ethnic group presented its own problems. We at the college were motivated by two things: to give our own students an opportunity to learn from foreign students, and to train the foreign students so that they could go back and help their fellow countrymen. In a shuffle of this sort a lot of things often go wrong. It's one thing to teach them to become biologists or engineers or accountants so that they can be active agents for technological developments back in their own country. It's quite another thing to Americanize these students so much that they may lose contact with their native surroundings.

In this case, the woman's daughter was a student at our Rutherford campus. Although we were giving her free tuition, someone had to provide for her transportation to America, room and board, and expenses for books and for incidentals. One thing I hadn't realized was the problem of clothes. In Sierra Leone, the women wear just a light dress throughout the year. In the often very cold months in Rutherford, such a dress would clearly be impossible. Happily, it was easy enough to get people to contribute proper clothing.

A clothing problem also came up with three Cuban students who had fled from Castro and who had among them only one jacket and one sweater. I made an appeal at the next Rotary meeting. The Cubans were strapping young men. I happened to be president of the club that year. I asked all members five feet nine or over to stand. I told them of the plight of the Cuban exiles. I added, "I'd like you to rummage among your clothes and donate suits and especially overcoats." The students wound up with enough clothes to outfit a whole squad.

On another occasion I had a Nigerian student helping me on a television series I had organized called "New Nations." He was a journalism major and the experience was invaluable to him. He would come to my home every Saturday morning and would drive with Sally and me to Channel 13 in Newark. I noticed that his clothes were shabby. I gently asked him whether he would like some of my suits and he replied eagerly in the affirmative. I took three of my suits and spent $32 to have them altered to fit him. Some weeks later, after noticing that he was still wearing the same shabby clothing, I asked him how my suits fitted him. He seemed embarrassed but finally he admitted that he had sold them to a second-hand clothing dealer. And that taught me another lesson. We forget that a foreign student needs pin money for odds and ends—a haircut, a package of cigarettes, laundry, a movie. The scholarship we give him covers only a portion of his needs.

Another problem is that of food. Usually, in the home country, students eat what is close to the soil, usually simple organic food. Transplanted to America they cannot get the food of their country and adopt all the worst nutritional habits of the United States: frankfurters, coke, pie, ice cream. The Sierra Leone girl was no exception. As she carried her books on her head from the nearby dormitory to her classrooms she couldn't understand why her classmates thought it was strange. This was the way one carried things back home. It was awkward to carry things under her arm. Her way made sense to her. It encouraged good posture and a strong back, which many Americans lack. She enjoyed the American food immensely, but she did not enjoy the same health she did back home.

Back home, schooling was mainly by lecturing to passive students. Here she was shy about entering into discussion because her English had strange pronunciations and inflections. Besides, she felt she didn't know as much as her fellow students. In spite of the fact that we had put her in a small dormitory with about twenty students, she felt lonely and homesick. In later years, she told me she used to sob to herself far into the night. Probably she was better off with us than in many other institutions. Rutherford is a small town and yet it is only twenty minutes away from New York. Our students were genuinely interested in helping her and did. But they too had their lives to live, boy friends to see, families to see, cars to ride in. There was a chasm, and the foreign girl felt it. Little by little it became narrower and, by graduation time, she learned to hold her own. She had good native intelligence and learned to participate in American class discussions. Her instructors were all sympathetic to her and she began to realize her own strength.

The student from Nigeria had caused me some anxious moments a few months before. I was accustomed to call the college switchboard each evening, no matter where I was. This particular night I was attending the symphony. Before the concert started, I called the college and the operator asked me to call a Newark police captain. When I spoke to him, he told me that they had one of our foreign colored students locked up. I asked where he was from. "Congo, I think."

I knew we didn't have any students from the Congo so I asked the captain to ask the student whether he was from Nigeria. Sure enough, he was. What had happened?

"Well, he got into a fight in a restaurant and we had to lock him up."

"Look," I said, "I don't want that student kept in jail. I will be responsible for him. I will send one of my deans to pick him up or I will come myself. I should appreciate it if you will tell him this and have him wait in an office rather than in a cell." He acquiesced. I immediately called Dean Clair Black and asked him to hurry to Newark, sign out the student, and accompany him home.

The next day I had the student come to see me and asked him what had happened. He explained he had stopped in a restaurant in Newark. He saw some friends and went to get an extra chair for himself. A lady was sitting at a table for two. He had thoughtlessly taken the chair without asking for permission. Unluckily her husband came in and, in a flash, harsh words led to a fight. I explained to the student that one doesn't take a chair away from a table without asking the person or persons sitting at that table and, no matter how sensible it may seem, if a person denotes displeasure, the chair must be left. He admitted that he was wrong and that he had been hotheaded. But I was glad that that student hadn't spent a night in an American jail. It would have engendered hatred for America for life.

To get back to the Sierra Leone mother. The three dollars she had saved for her daughter represented three weeks of work picking fruit from trees in the outskirts of the town and selling it in town. The girl was a bright youngster. We had supplied a tuition scholarship. The tribe, through some obscure governmental arrangement, had provided transportation and money for room and board. The family had to take care of everything else. But what to us was an insignificant amount was to them the savings of a lifetime. When we bring a foreign student into an American college, we have a responsibility of the highest moral order. We are uprooting him, and, unless we are prepared to assume his orderly and rational guidance in every detail, we have no right to bring him to America. We found this to be a fact in practically every emerging country we visited: fourteen, to be exact.

We probably encourage more foreign students to come to America than we have a right to. Many of ours came with grave intellectual as well as material deficiencies. First of all, many lack skills in reading, writing, and speaking English. Therefore remedial work should be provided for them. I had tapes of English courses made available for all foreign students because even those who speak intelligible English in their home country usually have a different pronunciation and inflection, which makes their speech sometimes difficult to grasp over here.

Our academic calendar is also very hard for foreign students. There are vacation periods at Christmas, at the end of the first semester, and at Easter, when in some cases food service stops and the dormitories are closed. The American student goes home. What does the foreign student do? He is lost and lonely. Then there is a long summer vacation. The foreign student legally may not work. If he has money, which he usually doesn't, he can travel and see the country. Here and there, through the generosity of some school friends or of the institution, help is given him. But once you realize how insufficient this help is, it is amazing that there aren't more students who become embittered in spite of our noble intentions toward them. When a foreign student comes to an American college it would be better if he studied right through the year and were thus enabled to finish his course of study earlier. But then, I would make sure that he is encouraged to go back home. Keeping them in this country too long draws them too far from their home surroundings. If a student from a poor or emerging foreign country comes to the United States and is deeply attracted to our style of living and our luxuries but is then plunged back into his native land, the result can be a traumatic one. I know of one foreign student who was so completely Americanized that when she went back she was literally a foreigner to her own people. She finally was able to come back to America, where she still is. Sometimes, and this is probably occasionally desirable, the student may refuse to return to his country, deciding to stay in America. This, of course, nullifies the entire purpose of the foreign student program.

On the basis of our experiences, we decided to have fewer undergraduate students and more students who wanted to study for their master's degree. For instance, we had some students who, in spite of everything we could do to help them, simply could not make the grade. A student who goes to a foreign country and fails simply cannot go back to his own people. Some, no matter how much remedial work we did with them, could never master English. The master's students are more mature and have a clearer professional goal in mind, and they usually are more fluent in English. They usually finish their work in a year or a bit longer and the whole process makes for a more co-ordinated, more consistent, and more efficient operation.

12

Honorary Doctor and Indian Chief

"A call from the State Department for you, Dr. Sammartino." I picked up the phone and an official at the other end said: "We know of your interest in African nations. The president of Upper Volta is here. Would it be at all possible to extend an honorary degree to him?"

This was Thursday. He would be here in only five days.

Hesitatingly I replied that this presented problems, that it isn't easy to have an honorary degree voted by the Board of Trustees; certainly not this quickly. I suggested that perhaps we could arrange to have him made an honorary Indian chief. This I could probably arrange in a few days. I knew the Chief of the tribe and as a matter of fact we had helped the tribe in some of the research of the ritual. I felt reasonably sure that if I telephoned him he would quickly make the arrangements for such an outstanding guest to become a honorary member of the Shinnecocks, who had their reservation on the south shore of Long Island near Southampton. When the explorers' ship neared the Narrows, the camp fires of the Shinnecocks were probably the first signs of life the earliest explorers saw.

The caller replied that he would call me back. Within two hours he called back and said, "We like your idea about the Indian ceremony and we'd like you to do it as well as the conferral of the degree."

I replied weakly that I would contact all the trustees and poll them. It was an unusual procedure. Usually it took the better part of a year for a nominee for an honorary degree to get final clearance. Along the way, there would often be ninety percent mortality among the nominees. Over the first twenty-five years of Fairleigh Dickinson's exis-

91

tence the average was about three honorary degrees a year and we were proud of all of those we had given out. Very few had been big donors and most of the big donors had not been given honorary degrees. Generally the recipients were people who had served their fellow citizens well.

Polling by telephone was a new procedure. At any rate, I started to call the trustees and was able to reach all of them, even though two were out of the state. Unexpectedly, they all reacted favorably. Now I was faced with two problems that had to be solved in exactly three days: to get the Shinnecock tribe to agree to the honorary chief ceremony and to get them to arrange it in the time available, and to make arrangements for the conferring of the Doctor of Laws degree. Since President Yameogo was staying at the Waldorf Towers, I called my friend, Clyde Harris, who was managing the Waldorf-Astoria and asked his help in providing a place for the Indian ceremony. This he did very willingly. I decided to have the degree ceremony at our Madison Campus only because such functions hitherto had been held at our other two campuses, and I thought it was Madison's turn.

It was now Friday, one day after the call from the State Department. I started to work immediately on getting a flag of Upper Volta and a copy of the anthem. Neither was available. Finally I got a small flag from the Upper Volta Embassy. I called the State Department for a copy of the anthem. They rushed over by special messenger both the music, which had a tricky rhythm, and a tape recording of the Marine band rendition, since President Yameogo had already been received at the White House. Then I called my friend Estelle Liebling and told her I wanted a tenor to sing the anthem. She got Vernon Shinall of the Metropolitan National Opera Company. The anthem had French words; Mr. Shinall knew little French, so I had him come over to Rutherford the next day to teach him the words. By that time the messenger had arrived with the music. First I taught the pronunciation of the words to the artist. Then we both listened to the Marine Band tape and I played the accompaniment. If anyone had told me that I would be a music coach, I would have laughed, but here I was doing exactly that. After two hours, the artist knew the anthem well enough and had, in fact, become intrigued with the unusual beat of the anthem.

In the meantime I was working away with the Shinnecock Indians at the Southampton Reservation. I finally reached Chief Thunderbird, who was out on the golf course. I succeeded in convincing him and his colleagues of the importance of the occasion, and that weekend the squaws put together three beautiful headdresses—one for the President and the other two for his two sons.

In the meantime, I was also working with Dr. Sam Pratt, Dean of the Madison campus. Those faculty members and students who were free were invited to the ceremony, which was scheduled for ten o'clock Monday morning. The presidential cortege was an hour late, which gave me a chance to tell the students a little about Upper Volta, how it had achieved its freedom, how it got along economically, and its relation to other African states. Finally the President arrived, followed by his ministers and representatives of the State Department, including an official interpreter, since the official language of Upper Volta is French.

I gave my speech in French slowly, sentence by sentence, for the State Department interpreter to translate it into English. I compared the problems of Upper Volta to some of those we had in our own country when we were so small and had so little. I said that it is only through the study of small and new nations that we can delineate the problems of any nation. Nothing could seem more dissimilar than Upper Volta and the United States and yet, when we began, we had many of the same problems: division among ourselves, the role of agriculture, the encouragement of industry, and the development of education. Years before, I had asked our social studies department to include a unit on Israel because it afforded such a wonderful opportunity to study the elements that make up a nation. Mr. Shinall sang the Upper Volta National Anthem and there were tears in the eyes of President Yameogo. It was the first time outside his own country that anyone had taken the trouble to sing their national anthem correctly and with the proper rhythm. Even at the White House this hadn't happened. At the end of the ceremony, a number of ministers came up to me and said in French, "You know, this is the first ceremony where we really understood everything and enjoyed it." I was glad that Fairleigh Dickinson had participated in this historic occasion—a real first.

That evening at five, the Indian ceremony at the Waldorf-Astoria took place. In the official State Department schedule we had exactly thirty minutes for the ceremony. Indian lore is a fascinating subject for foreigners. To be made an Indian Chief is one of the greatest accolades one can receive. The representatives of the Shinnecock tribe were there in person and were all eager to play their roles. This was the first time they had welcomed into their tribe the head of a foreign nation. They chose the name of Shagwannuck, meaning the Good Dweller of the Land, for President Yameogo. At the end of the ceremony, he came to me and said,

"Could I buy headdresses for my two boys?"

I had already placed the headdresses on the table for extra decoration. It was nice to say, "I have already ordered two for your boys." He smiled most appreciatively.

I had a chance, even though it was a fleeting one, to converse with the President at our two receptions and later at other receptions. He was an unassuming man who was dedicated to the growth and development of his country. He seemed to be the kind of person who, while maintaining the folk ways of a small, tribal nation, could successfully bring it up in the industrial world so that it could earn enough money to buy the things it needed to raise its low average income. Like the Ivory Coast, Senegal, and other former French colonies, Upper Volta had opted to keep up its cooperative relationship with France. Some months later, we received through the Upper Volta embassy a package containing the longest tablecloth I have ever seen—about twenty feet in length, hand embroidered in native design. We have used it doubled over on state occasions and it still almost touches the floor. By the time our letter of thanks reached him, there had been a palace revolution in his country and President Yameogo had been put in protective custody and later in prison, where he tried to commit suicide by cutting his wrists. At a United Nations reception I asked about him, but I got nothing but anguished protestations of ignorance. And that's how one of our honorary alumni ended up in prison!

Since I had to keep refreshments at a minimum in view of the brief time allotted, I thought of firewater as the most appropriate drink and I felt that American bourbon was about the closest to it that I could order. I gave my instructions to the waiter: Have fifty old-fashioned glasses, each with two cubes of ice and one jigger of bourbon. Pass it around as soon as I give the signal. I thought I had made my instructions as simple and as explicit as possible. What did the waiter do? He began to ask each guest how they wanted the bourbon, with water, with soda, on the rocks? Some of the ministers who couldn't understand English were visibly confused. One answered "Champagne." In five minutes the waiter succeeded in completely messing up the end of the ceremony. Most guests, happily, had to hurry away without sampling "firewater."

Then occurred one of the most nightmarish events that could ever befall a wife. One of the guests, Virginia Pollak, invited those of us who were left to dinner at the Siro restaurant, which was a few blocks away. In a mix-up of cars, I went away thinking that Sally had gone with the Pollaks. In the meantime Sally was upstairs, making sure that we had not left anything behind. The waiter approached her with a full bottle of bourbon, out of which only one drink had been poured. He suggested that she take it along, since it had to be paid for and it was a shame to waste it. Unwisely, Sally took it, hid it under her coat, and figured she could easily get to our waiting car. By the time she got down to the 49th Street exit of the Waldorf, I wasn't there and she didn't know at what restaurant the dinner was to be. She hailed a cab and figured she might spot me or our car or perhaps the Pollaks. One does not realize what a big and confusing area just a few blocks of New York can be. After fifteen minutes of vain cruising she gave up. But now she realized she had changed bags at home and had no money. In disgust the taxicab driver took her back to the Waldorf. By this time she was sobbing and was a nervous wreck. She went to the banquet department and asked for a loan to be charged to my account. The lady on duty espied the hidden bottle and must have thought the hysterical lady was surely an alcoholic. After much talking, the assistant manager finally gave her five dollars. In the meantime, at the restaurant, I became more fidgety by the minute. I went back to the Waldorf, scoured it from top to bottom—no Sally. We were supposed to go to the opera that night. She had the tickets. I explained my plight at the subscription office of the Metropolitan and paced nervously in front of the section where we had our seats. Finally at a quarter to eight Sally came in, completely beaten. We rushed into each others' arms and, of course, neither of us was in any condition to enjoy the opera. The people around us looked at us pityingly, not knowing what to make of this hysterical scene. We returned home to Rutherford and it took us at least two hours to resume our serenity.

A few weeks later, Sally and I were dinner guests at the White House. As we went through the receiving line, the then Chief of

Protocol, Mr. Lloyd N. Hand, presented us to President Johnson, who knew us since he had spoken at the Teaneck Campus of our University, and said, "Dr. Sammartino has been very helpful in entertaining President Yameogo of Upper Volta." I had brought along three colored photographs of the Indian ceremony and showed them to the President. He put on his spectacles and looked at the three photographs carefully while I explained how we had made him an Indian Chief. President Johnson had had an Indian show at the White House, but no chief's ceremony. He turned to the Chief of Protocol and said. "Isn't that wonderful! Why didn't we think of making him a chief?" It was a diverting experience and it made us feel that all our efforts had not been in vain.

One of the strangest requests ever made in academic circles was in our School of Dentistry. A required course in dentistry is human anatomy, which involves the dissection, bit by bit, of cadavers. When we opened the school, bearing in mind the terrible stench of the formaldehyde that always permeates dental buildings, I decided to build ours away from the main cluster of buildings. We happened to have a surplus building from the old Camp Shanks, and we built into it a good refrigerator unit where we could store twenty-four cadavers. We couldn't get the cadavers in New Jersey so we used to buy them in New York. They were usually bodies of derelicts found in the streets, with no claimants. We paid the city about thirty dollars for each.

A dissection laboratory has a number of stone or metal slabs upon which the body is placed. In a school of medicine, usually two students share one body. In a school of dentistry, since the students are chiefly interested in the anatomy of the head, there is less need for as minute a study of the entire body, so that in our school four students shared one corpse.

Let me digress a moment to schools of medicine. I have been in medical schools in European countries where sixteen students worked on one body; in some Asian countries, even more shared a body. Many of these foreign students, when they graduate, do their interning in American hospitals because of the great need for hospital personnel. What is happening, of course, is that medical graduates with very imperfect training are learning the medical trade on a sort of apprentice method, but heavens knows at what expense to the patients.

At any rate, in our laboratory we had the best and most modern equipment: beautiful stainless steel covers for the corpses, interesting scientific pictures on the walls, and absolute cleanliness. The place was almost odorless. A truck with the bodies backed up to a platform near the refrigerator and the gruesome task of hanging up the corpses by tongs pierced into their ears, was expeditiously performed. Then, as they were needed, the cleaner would put them on the shining plat- forms. The students whittled away, dropping the flesh into pails which were emptied daily. The remaining skeleton, or some part of it, was used for instructional purposes, but if not, the skeleton had to be signed in to the city authorities and an official receipt procured. After all, you can't throw away the remains of a human body without some sort of explanation. Incidentally, every once in a while, after going through the expense of having a corpse brought all the way from New York, a distant relative would somehow show up and claim the body, in which case we would have to go through the double expense of sending it back.

A human anatomy laboratory generally costs about $200,000, if you figure the proportionate cost of the building. Ours cost only about $5,000, and that included the cost of the surplus building. It was the most efficient, the neatest, and the best-looking one I have ever seen. Whenever I showed a visitor around the campus, male or female, I always included the laboratory, to show what could be done inexpensively but with maximum educational results. I always made sure it was a quick visit, for I was apprehensive that the visitor might faint. Ralph, the cleaner, was always very cooperative. As we breezed through the laboratory, we would finally come to the refrigerator. Ralph would open it for a few seconds, just about enough time to make the visitor realize the final result of all life. I noticed that all the male cadavers had penises sticking out about a foot. This usually pro- duced a humorous remark if the visitor was a male. I don't know what the female visitors thought. I wrote a memo to Dean Wilson and asked him if it were possible to reduce the sizes of the penises. He looked into the matter and replied that it indeed would be possible. What had happened was the following. The corpse, when bought, would first be sent to a school of embalming. The students, as a matter of horseplay, would pump extra formaldehyde into the penis and thus the contretemps occurred.

A female member of the Board who was an artist made a sketch from memory of the cadavers in the refrigerator and sent it to me on the next day after her visit, with the notation: "The memory of the cadavers has been haunting me all night."

Actually, the decision to establish a dental school was made at the first graduation of our dental hygiene school. Some of the State Dental Board people were there and, by pure chance, my father-in-law, Mr. Scaramelli, asked one of them why we couldn't have a dental school. He saw no reason why we shouldn't. This was all the urging I needed. I immediately set up a committee of dentists to suggest a curriculum. After four months, the group was still haggling about the make-up of the course of study. I then asked my Director of Libraries, Dr. George Nelson, whose specialty had been biology, to make a study of dental school curricula and suggest to me an ideal pattern. In three days he had a suggested curriculum, which we then submitted to the committee, and they adopted it practically in toto. They then proceeded to lay plans for the laboratories needed. Dr. Walter A. Wilson came on board as the dean and, in 1956, had our first group under way. I made three conditions. First, I wanted the clinical work started as soon as possible so that the students would have maximum practice before graduation. Second, I wanted nutrition to be an integral part of the curriculum, because a dentist can usually spot malnutrition long before a patient gets to a physician. Third, I wanted special reading training for our students because, in spite of the fact that all, without exception, were top students, reading faster meant better college work and later professional reading.

Now we have a new $9,000,000 dental building and the new anatomy laboratory has to be placed inside the building. Incidentally, we almost didn't get the building. We had the grant of $3,000,000 or so from the Federal government, but this had to be matched. I had arranged a meeting with then Governor Richard Hughes, bringing together members of our Board of Trustees and Board of Fellows. I made the point that it would make sense for the State to match the Federal grant because it would bring a much-needed facility into New Jersey. The Governor was sympathetic, but at that time there were so many rhubarbs in the educational field that nothing happened. Finally, in 1969, I was having a cavity filled in Dr. Giordano's office. Jim had been a member of the State Board of Dentistry, had helped to set up our School of Dentistry, and also taught the course on prosthodontics. He was tremendously interested in the success of our school. At any rate, after filling my mouth with gauze, he pointed the drill whirring at the rate of 250,000 revolutions per minute at me and said, "Peter what are you going to do regarding the new building?" I mumbled helplessly. I was no longer president and I really was in no position to exert pressure. After Dr. Giordano got through with me, he said, "Couldn't you call Mr. Dickinson (the chairman of the board of trustees)?" Meekly I said I would. I made an appointment to see him. When we met I told him that he was the only hope, that we needed $3,000,000 and had only $1,000,000 toward the project. We did not have time to raise the rest of the money since the Government deadline, twice stretched, would soon expire. Dick Dickinson made his decision quickly. "The School of Dentistry has to have this building if it is to retain its primacy. I'll give the $2,000,000." Later he stretched it to $2,300,000. And that is how the new building became a reality. I cite this as an example of how great decisions in educational centers are often made on the spur of the moment. It is a wonderful new building but, alas, the cadavers won't know the difference.

14
Heraldry

Just about the time we had devised our first coat-of-arms I realized that we had no mace for our academic processions. This was the war period and gas coupons were in effect. On one of my infrequent trips back from New York, just before the Lincoln Tunnel entrance, I gave a ride to an elderly gentleman who was thumbing his way into New Jersey. As we conversed, and as he told me he did gilding at the Metropolitan Museum of Art, I began to get an idea. Why couldn't he make a 16-inch coat-of-arms on wood? We could then attach it to a fluted pole about a yard long and we would have a mace of sorts! We stopped at my house for a few minutes. In one of the guest rooms, there was an old-fashioned bed with two fluted posts. It would make just the handle we needed. I sawed off the post, since we didn't like the bed anyway. The gentleman took it away and a week later he had completed the mace. It was beautiful, even though it was different. The only disadvantage it had was that it was light. A mace has to be heavy. In English ceremonials, one can almost see the mace bearer creaking under its heavy load. Immediately after the war I was in England and put in a request to various shops for a mace. I received a call from one. Yes, they had a mace for about twenty pounds —a real old mace. Here was a real bargain! I rushed down to the shop to examine my find and, to my disappointment, it was a small black affair about fifteen inches long. It was simply a rod of iron with a spiked ball at the end. This after all was a real old mace which originally served as nothing more than a blackjack. Whoever could wield a mace effectively had the power, and soon a mace became equated with power. It was long years afterwards that it became an embellished ceremonial object.

Years later a friend of the university, Mrs. Leo Pollak, bought us an original English silver mace which is now in use. It is one of the three English maces in the United States, the other two being at Yale and at the Norfolk Museum.

It was just about this time that we were evolving at the University the so-called horizontal and vertical deanships. We had three campuses and for each campus we had a dean—a vertical one. The horizontal deans held sway over their courses on the three campuses. For instance, Dean Herdman of the College of Education was responsible for the education courses on all three campuses—he was a horizonal dean. Which kind of dean took precedence in the hierarchical organization? This was the problem of protocol that affects government, business, and, in no less a manner, education, especially in a multi-campus university. One would be surprised at how sensitive administrators and faculty members are to their place in an academic procession which, incidentally, can be organized in one of two manners. Either the most important come first, with the fledgling instructors at the tail end, or vice versa, so that the procession winds up with the high and mighty trustees at the end. I have always used the first method. In order not to offend either group, I asked the campus deans to carry the mace ahead of the procession, much as flower girls precede the bride. I was intensely proud of all the deans and I didn't want to hurt any feelings.

Incidentally, when we dedicated our campus in England at Wroxton, I had Dean Haberly (horizontal) bring over the English mace. In that particular detail, we Americans were just as up on ceremony as the British. Here we changed protocol a little since he was mainly responsible for the curricular development at Wroxton. Photographers for the press and for television concentrated just on him. Overnight we had captured a little of pageantry in England.

I have always said that academic pomp is corn and I have always taken it with tongue in cheek. Whenever people are in academic processions they always seem to waddle rather than walk directly forward. The result is that these processions always take much more time than they need to all to the seemingly endless repetitions of "Pomp and Circumstance." In an effort to hurry up the enfilade, I had our bagpipe band play a lively march and, by Jove, it cut down the

time for commencement processions almost twenty minutes. The
marching affected not only the faculty but some five thousand
graduates.

Quite apart from our own activities, it is usual in academic circles
that, when a new college president is chosen, a formal induction ceremony is held. At that time it is usual for other institutions to send
formal congratulations to the newly-elected president. These may be in
the form of a letter or in some cases a hand-engrossed resolution
bearing the seal of the institution. I decided to do ours differently. I
had an imposing certificate in Latin printed on parchment, to be filled
in with the name of the new president. It read as follows:

UNIVERSITAS FAIRLEIGH-DICKINSONIANA
RUTHERFORDIAE
IN REPUBLICA NOVAE CAESAREAE

SALUTEM DICIT

JOHN JONES

LABORES RECTORIS PRAESIDISQUE RIGIDOS SUSCIPIENTI.
TE NE OFFICIA NUMMUM CONGERENDORUM OPEROSA,
NEVE SERIES INFINITA CENARUM GALLINACEARUM, NEVE
ACRES ALUMNORUM ADMONITUS PROHIBEANT, ACADEMICA
VITA DIUTURNA SANAQUE FRUI.

IN CUIUS SPEM LITTERIS HISCE UNIVERSITATIS
SIGILLO MUNITIS SUBSCRIPSIT.

PRAESES

But for those who could translate, the message was distinctly jocular,
hoping that neither the laborious duties of raising money, nor the endless rounds of chicken dinners, nor the bright suggestions of alumni
prevent you from enjoying a long and happy life. Many of the new
presidents hung the certificate in their offices. Some answered in Latin,
one soul even in Greek. To most visitors it probably looked like an
honorary degree. But to the new president it often provoked a wry
smile reminding him not to take temporal frustrations too seriously.
At many professional meetings the humorous certificate became the
bond for a new friendship.

Some of the young people today have taken a stand against pomp
and bourgeois ceremony. In doing so they want to substitute simple and

uncluttered meetings. I think they make a mistake. Life has to have color, ceremony, ritual—provided you don't become too stuffy about it. While I never attended my baccalaureate or master's degree graduation, I did attend the commencement at which I received my doctorate. As it happened, the heavens just opened up that day and the ceremony was in a shambles. At that point in life, it was important for me to engage in that bit of pageantry even though it was somewhat soggy.

When we first started the college, our coat-of-arms was very simple: a shield bearing a stylized version of the castle on a maroon background. The official heraldic description was: on a field gules, a castle argent.

Dean Loyd Haberly, whose hobby was heraldry, made the design. Since we tried our educational experiment with as much courage as we could muster but tried to have a good time while doing it, I suggested: "Fortiter et Suaviter," "Bravely and Pleasurably," which has become our motto. Some years later Admiral Rickover came over to the college to speak to our students. He happened to see the motto and asked what it meant. "Bravely and Pleasurably," I replied. "Well, that's a dumb motto!" "Why?" I asked perplexedly. "Because I've seen more men killed while they did things bravely than I care to remember. And if you try to have fun while you're doing things, nothing ever gets done." Well, that was that. But the motto still remains, and I like it. It isn't so stuffy as other mottos and it conveys the idea of our experimental turn of mind.

When we acquired Teaneck campus, we divided the shield in half, pushed the castle to the upper part and had a swan on the lower half on a wavy blue-and-white background. I chose the swan because of the pair of swans, mentioned earlier, that I had put into our Teaneck pond; I had to make the decision in a hurry and that was the best idea I could put forth. The wavy blue-and-white lines represented the Hackensack River. Corny? All coats-of-arms are corny. Incidentally, when we added our Wroxton campus in England, the Queen gave us a pair of royal swans. All swans except a few belonging to certain guilds belong to her.

When we added the Madison campus, the castle became a simple crenellation along the middle, the swan stayed put, and since Madison was then known as the Rose City, I had three Lancaster roses put at the top. Loyd Haberly patiently drew each new shield and composed the official heraldic description, which by this time was as follows:

> Azure; a fess gules crenellated; on the chief three roses argent; on the base a swan argent.

When we acquired our English campus we had a new problem. Luckily, through Loyd Haberly, I had come to know Michael Mac-Lagan, Senior Tutor at Trinity College of Oxford University. He was a great authority on heraldry. One evening we were having dinner with him in Oxford, and while we were eating, I brought up the subject of a new coat-of-arms for Wroxton.

"Michael, what should we include in it?" I asked. He took a paper napkin and suggested a crenellated shield to imply our Rutherford campus; a lion and fleurs-de-lys from the North family arms, since Wroxton Abbey was the ancestral home of the family; and three swans, to denote the swans at Teaneck. Later, the Queen gave us two swans so that the coat-of-arms could represent both campuses. No swans did greater duty internationally or academically. As a final touch, we decided to add three roses to represent our Madison campus, but made sure it was a Lancastrian rose to harmonize with the political leanings of the North family. The next day, I had the new coat-of-arms made up in stained glass, to fit in a window that had been broken for decades. But we weren't through yet.

We really could not properly use any coat-of-arms unless it was registered at the College of Arms in London. Here Ricky Davis of New College in Oxford helped us. The Heralds of the College of Arms are appointed by the Queen. They keep the official records and designs of all coats-of-arms from time immemorial. They must examine all new proposals, see that there is no conflict with existing arms, propose changes, and finally approve any new design. After many conferences, we agreed on the drawing. But now we had to get the permission of the Governor of New Jersey, then Richard J. Hughes. The final design for the coat-of-arms was really a lulu. It had some of my ideas, and also some of the Queen's Herald. Since Wroxton Abbey was the ancestral home of Lord North, the Prime Minister during the reign of George III, we had the fleur-de-lys and lion passant from the North coat-of-arms. A swan represents Teaneck, but now he is no longer the placid swan of the American shield but an angry, fluttering one who seems to be raising hell. A knight's helmet represents Ruther-

ford. A book to represent scholarly activities and a bee for energy are
thrown in for good measure. All of this is enveloped with a lot of
scroll work. The fancier the coat-of-arms, the more the College of
Arms has to charge you for the finished design. Finally the two end
leaves of the fleur-de-lys, which I have considered the most pleasant of
simple designs, became heads of angry swans. The description now
becomes:

> Per pale Azure and Or on a Chevron between two Swans rousant in
> chief and a Lion passant guardant in base three Fleur de Lys all inter-
> changed. And for the crest on a wreath of the Colours on a book fesswise
> Azure garnished a Bee volant Or as the same are in the margin hereof
> more plainly depicted. And by the authority aforesaid we do further grant
> and assign the following Device or Badge that is to say: A Fleur de Lys
> the two outer petals replaced by Swans' heads and necks erased Or

After three years, I finally got the official parchment duly attested to
by:

Garter, Clarenceux, and Norroy and Ulster, Kings of Arms, whose
learge sealing-wax imprints lend a royal and imposing cachet. It was
fun through it all. I'm glad I didn't waste too much time, but whatever
hours I did spend were pleasurable. Fortiter et Suaviter!

15
African Drumbeats

I was standing on the massive steps of the new building of the University of Abidjan in the Ivory Coast waiting for the President of the Republic to arrive. All of a sudden, I felt my trousers slipping beneath my academic robes. I had forgotten to put on a belt! The Presidential cortege was just pulling in. The red-trousered guard of honor burst forth with a fanfare of trumpets and a staccato roll of drums against a crescendo of tom-toms beaten by hundreds of native drummers. About twenty thousand tribal natives had come in from the neighboring towns for the great event, and their voices and eerie whistles filled the enormous square. One of the great moments of my life and my pants were slipping!

With my left hand pressed tightly against the top of my trousers I walked down the steps and shook hands with the President and escorted him to the hall within the building. On behalf of Fairleigh Dickinson University I was conferring upon him the honorary degree of Doctor of Laws. As the emerging countries of Africa were assuming importance in the family of nations, I had proposed to the trustees that we honor three of the African Presidents: President Senghor of Senegal, President Houphouët-Boigny of the Ivory Coast, and President Tubman of Liberia. The last had received his degree while on a State visit to the United States. President Senghor was to receive his while attending a projected Afro-American conference in America, but the conference was postponed a number of times and finally dropped. When I was asked to give a commencement talk at the University of Liberia, I took the occasion of visiting nearby Ivory Coast to present the degree to its President.

I therefore made my arrangements to step off at Abidjan so that I could call upon the President and present him with the honorary doctorate which our Trustees had voted at my recommendation. I was traveling in territory unknown to me. Although we had visited countries in North Africa, this was our first trip to West Africa; as is natural, we were somewhat apprehensive as to how things would work out.

On arriving at the airport in Abidjan, I had felt completely lost. There was such a hodge-podge assemblage of people, all gesticulating and many yelling and pushing their way through the milling crowd. Finally, as I wended my way through the customs, a grinning, effusive native accosted me. Over the years, I have been approached by a great many hucksters in strange lands who are ready to supply one with anything: hotels, automobiles, women, local currency, gifts, you name it. My natural inclination is to shove them off, which I did in this case as I started to look for a taxi. Again he accosted me, and this time he showed me a crumpled piece of paper which he attempted to straighten out so that I could read it. All I could read was "-ARTIN-." Half curious, I followed him and he led me to a glistening car which had a government shield. By this time I could make out some of his French patois. I heard "votre chauffeur" and "Ministre." I decided to take a chance and get into the car, making sure that he heard "Hotel Ivoire, Hotel Ivoire." He nodded gleefully "Oui, 'sieu," and off we went at breakneck speed, with the horn sounding off at every possible opportunity. We arrived at the hotel, which is plush, sophisticated, and full of hubbub. We were ushered in and, to my surprise, a suite was waiting for us. I didn't want an expensive suite. I had ordered a simple twin-bedded room and bath. "But Monsieur, the Minister insists we give you a suite. You are the guest of the Government." Then the truth dawned upon me. The Government was rolling out the red carpet for us. We were asked to pay a call at the Ministry of Education and the plans for the actual conferral were set. It was suggested that it be at the newest building of the University of Abidjan, to which I assented only too willingly. I then dropped by the United States Embassy. Fortunately, the Ambassador was in and received me. "You're creating quite a stir, and it's all good," he said pleasantly.

We had an extremely interesting chat on developments in West Africa, then Sally and I went to have lunch with the Ivory Coast ambassador to Washington, who happened to be back in his native land.

As we looked at the menu in the restaurant, I happened to say that I wanted to eat very lightly.

"That's right," he said; "tonight's the big shindig." He was very proud of his Americanisms.

"What's tonight?" I asked innocently.

"Don't you know?"

"No, I don't."

"What? Don't you know the President is giving a State dinner in your honor?" Well, this was news to me. Somewhere in the complicated wheels of bureaucracy, someone had forgotten to invite us to our own party. I could see that the Ambassador was upset over the lack of communication. By the time we got back to the hotel, evidently the proper wheels had been set in motion and there were frantic calls from one of the ministers. He came over to the hotel personally to extend the formal invitation for the evening banquet.

"Well," I happened to say, "Isn't that wonderful! Today happens to be our wedding anniversary and my wife's birthday. This will be a wonderful celebration!"

Later on I was to find out that the President had personally requested that two dozen red roses be flown from Paris for Sally for the banquet. But when he found out it was her birthday, he had a beautiful gold brooch delivered to the hotel with a personal message. Life is just as strange in diplomatic channels as it is in education. The top man has consideration and thoughtfulness to follow through on a personal detail and somewhere down the line some third- or fourth-echelon person is too unimaginative to take care of an important detail quickly.

At any rate, that afternoon the time approached for the formal ceremony. My ever-grinning native chauffeur drove me to the University hall and Sally and I put on our robes.

Now, with my sagging trousers I made it to the anteroom and, while the notables were being seated, the President was escorted into the other anteroom so that he could put on a black academic robe and wait for the solemn entry. I took the occasion to hitch up my trousers and my academic dignity seemed to be safe for the moment. We entered the hall where the diplomatic corps of practically all the nations represented in Ivory Coast were present. It was the first time in the short history of this new nation that an affair of this sort had been held. It was a great occasion for the academic hierarchy to enjoy the full honors of their profession. In a sense history was being made. An affair such as this at at our own White House would have had every minute accounted for. Everybody involved would have received a set of mimeographed sheets stating exactly when to step forward, when to speak, how long, when to shake hands, when to leave, and a hundred such details. Here, the chief protocol officer and I improvised from moment to moment. It was a new experience for him; it was new for me. But I have never let protocol frighten me. I have never hesitated to make up my own ritual and to have fun doing it. Of all the forms of protocol, the academic variety comes pretty near being the corniest. You might just as well take it tongue in cheek, make it as enjoyable as possible. At any rate, probably unconsciously suggested by the Catholic mass ritual, I had devised in awarding honorary degrees the idea of having the recipient clothed in ordinary black academic robes. At the proper moment he is divested of these robes and clothed in the brilliant red robes of the honorary doctorate. The doctoral hood is then draped over his head and, as a final fillip, a special *collier,* which I designed and which consists of a six-pointed medallion attached to a maroon and white ribbon, is tied around the neck. So it was on the present occasion, with President Houphouët-Boigny. Finally, I read in Latin the diploma itself and handed it to him. All of this took place on the stage, with Sally assisting me as a sort of academic acolyte. When it was all through, the audience burst out in genuine and hearty applause.

I spoke in French and I compared the President of Ivory Coast to George Washington in our own country. But I also pointed out that George Washington didn't have to worry about many of the problems that are presented to the President of a new country: television and radio, automobiles, immediate creation of new elementary and secondary schools, airplanes. What we in American had achieved slowly and over a period of two hundred years, a new country was supposed to create overnight. I am sure that whatever I said must have pleased President Houphouët-Boigny because in his response after he received the degree, he was very warmhearted in his praise of the United States and of Fairleigh Dickinson University. He accepted the degree graciously and gave every indication of being appreciative of the fact that someone outside his own country had taken the trouble to honor him for the work he was doing in building his country. The ceremony ended, we marched off with full academic dignity. I escorted the President back to his waiting limousine through lines of colorful honor guards, and as he came out there was a wild outburst of native cheers with a deafening beating of drums. It was a day worth remembering. Throughout all of this I still had my left hand pressing against my trousers. Finally I got back to the anteroom and hitched up my trousers again, happy that my academic dignity was still intact.

We had barely time to dress for the formal state dinner at the Presidential Palace, which, incidentally, is much bigger and more ornate than our White House. It had been presented to Ivory Coast by President De Gaulle. As we entered, the President and Mrs. Houphouët-Boigny were there to greet us and to present us with a beautiful photographic album which, incredibly, had been put together in the few hours between the end of the afternoon ceremony and the evening function—another example of thoughtfulness at high level. Later he was to send me a set of color photographs which, naturally, could not be made ready for the evening. Sally thanked them for the brooch and the flowers. Photographs, a brooch, all the mementos that we cherish in our twilight years—aren't such "remembrances of things past" part of our lives for all of us? At the enormous horseshoe table in the glittering dining room, the gold service was out. I sat at the right of the President, Sally at his left. The President was a physician and we talked in French about health and nutrition. There was a great big basket of luscious fruit before us. I remember his taking up one of the very small native bananas, about five inches in length, and saying, "This is the most nutrious and the tastiest banana in Africa." It was a sumptuous dinner in the most sophisticated French manner, with two wines and champagne at the end. Toward the end of the dinner the President arose and offered a toast to Fairleigh Dickinson University and to us. I returned the toast to Ivory Coast and to President and Madame Houphouët-Boigny. I have been to three formal dinners at the White House. This was much more dazzling, much more formal, and yet fully as sincere.

The next day we went to visit one of the Peace Corps volunteers who was working in the small village of Cocadi, not too far from Abidjan. The newly developing African countries are so new that not far from glittering buildings and villas of the capital stand the tribal villages that have not changed in centuries. Throughout our travels, even before I had been appointed to the National Advisory Committee of the Peace Corps by President Johnson, I had made it a practice to visit Peace Corps installations whenever the opportunity occurred. In some cases we went many miles out of the way to touch base with one of our own students who might be serving in the Corps. We were one of the first colleges to encourage visits by recruiting representatives of the Peace Corps and to do it on the three campuses. We decided quickly to give credit for such off-campus experience long before it became fashionable to do so. Whenever I met Peace Corps volunteers I always made it a practice to write to their parents and let them know what their son or daughter was doing. The little village we visited consisted of not much more than a few hundred, mostly mud, houses and a few more imposing structures along the main street which, incidentally, was a dusty dirt road. The young lady volunteer was a black girl from Atlanta and was doing what I thought was a marvelous job of helping young mothers in the village. Young mothers means girls of twelve or thirteen. She was teaching them children's dances—not for their children, but for themselves. As we passed one hut, one such girl-mother, in her best Sunday dress and carrying a fancy pocketbook, was embarking upon social calls. She was a married lady now and protocol required such formal visits to the homes of her classmates. Later we went to the "école maternelle," where our volunteer was guiding the young mothers. That particular day they were weighing the babies. As one baby might show a gain of a few ounces in weight, shrieks of delight would emanate from the proud mother. Later we visited with the volunteer at her apartment. It was a fairly good apartment, five rooms or so, probably one of the best in town and probably better than her own room in Georgia. She was in seventh heaven. She felt wanted. She loved her charges. She enjoyed the freedom and independence of her living quarters. I can truthfully say that everywhere I have seen Peace Corps volunteers at work I have been impressed by their desire to serve, often under very primitive conditions. I have found them humble, imaginative, and hard workers. Here and there, I imagine, along with their strenuous schedules, they found time for amorous extra-curricular activities with other volunteers, but I see nothing wrong with this and indeed I have witnessed many wonderful marriages that have developed from Peace Corps stints. Sally and I even spent a few days in the mountains of Puerto Rico where Peace Corps volunteers are trained under rigorous conditions. We slept in primitive huts, brushed our teeth in outdoor showers, and washed our dishes in scalding hot water just so that we would know how Peace Corps people are prepared. We saw experimental chicken farms for training in St. Croix and barrio centers in Colombia that were simply the front room of the volunteer's living quarters. In Thailand we were able to have discussions with those volunteers who could travel to Bangkok and tell us of their problems. We found it thrilling everywhere we went. There was a spirit of service, of dedication. In some cases the volunteer functioned as an English teacher in the local lycée in French-speaking countries. Here he might not be able to function so fully or so creatively as he or she wanted, but still the volunteer was serving a need.

From Ivory Coast we went to Dakar in Senegal. Although the American Ambassador, Mr. James H. Wine, there had invited us to stay with him, we had elected to stay in a hotel. We reached the hotel at about one in the morning. The room supposedly reserved for us was still occupied and no other room was available. To make things worse, there was a power failure and the whole hotel was plunged in darkness. Luckily, each one of us was given a small candle. We decided to make the best of it and to stretch out in the lobby. As we were about to go to sleep, a young German lady approached us and in a frantic, fearful voice asked us if she could stay with us.

"Of course," we said.

"I want a picture of this," I said. "This is the only time in my life that I am going to sleep with two ladies in the same room."

I still have a snapshot of that awful night. Can you imagine yourself in a strange hotel, in almost total darkness, not knowing where anything is? And can you imagine trying to find the washroom, feeling your way in a dark lobby, trying to find the door with the magic "Messieurs" or "Dames" for Sally and our new-found friend. And, having found "Hommes," trying to shave in the early morning by the flicker of one four-inch candle? All nights come to an end. As the light began to filter in, our spirits rose and we would burst into laughter at the ludicrous situation. At this point I decided to accept the invitation of the Ambassador. I called the embassy; the ambassador had returned to Washington but his deputy, Mr. Donald Easum knew of my arrival and invited us to be guests at the Embassy. He sent a car for us and soon we were safely ensconced in the American compound. Mr. Easum arranged a meeting with President Senghor of

Senegal, and at this meeting we made tentative arrangements for him to receive the honorary doctorate which had been voted by the trustees. This was to be on the occasion of the African Cultural conference to be held somewhere in the United States. As I have stated, this conference never came about, and somewhere the doctoral robes ordered for President Senghor are still floating around. Too bad, because he is one of the few African leaders who has a real knowledge of Africanism and in addition has a thorough immersion in Western culture. He is a poet in his own right and, in fact, we printed some of his poetry in the Africa issue of the *Literary Quarterly,* which was established at Fairleigh Dickinson University on the insistent prodding of Professor Charles Angoff of our English Department.

Our trip to Africa had a strange effect on us. Years before, I had arranged for the majority of our Social Science department to visit Africa at college expense. I wanted them to have an orientation in three areas: West Africa, Middle East, and Russia. I don't know of any other institution that has done this. The members had their fare and all expenses paid. The idea was to get an orientation in these areas so that they could bring their knowledge into the bloodstream of the institution in which they were teaching. We did the same thing for Southeast Asia, and I am convinced that the money was well spent in terms of value to the students. I am happy to note that many of the professors went back on their own and enlarged their sphere of knowledge of many parts of the world.

It was just about this time that, during our summer stay in Maine where we rented a house from a well-known Rutherford lawyer, Mr. Milton Ignatius, I got another idea. Milton had bought a large estate not far from Treasure Island, off Sorrento in Maine. Among the houses on the estate was a large log lodge that contained an immense living room, a bedroom, a kitchen, and a bathroom. Milton didn't know what to do with the lodge. I suggested he give it to the University. I told him that if he did, I would invite ambassadors from the United Nations. At first, he demurred, but on August 15 of 1961 he came to me and said: "Peter, I know this is your birthday, and I want to give you a birthday present. I'll be glad to give the lodge to the University as you have suggested."

The next year we began to have United Nations ambassadors use the lodge, one week at a time. The first one to use it was Louis Ignacio-Pinto of Dahomey, who has since become a justice of the International Court at the Hague. Some of these ambassadors have lectured at Fairleigh Dickinson University or have taught courses. Many have become lifelong friends. All in all, there are few ambassadors at the United Nations who do not know of the concern at our institution for international well-being and understanding.

16

Student Freedom

I was able to create some freedom for exceptional students in those happy days when college presidents had some time for educational experimenting and could try out new methods without going through ten faculty committees or incurring the wrath of vested interests. Colleges, like high schools and elementary schools, eventually wind up as a wholesale arrangement of machine-like schedules, with every student put through essentially the same curricular mold. I asked two girl students whether they would not like to free themselves and work out their own educational objectives. I would act as their mentor and they could see me once a week, but if at times they wanted to call me by telephone or write to me I would always be available to guide them and to answer their questions. They could take any courses they wished and have unlimited opportunity for independent reading. If there were any books they wished placed in the library we would have them ordered, within reason. They could engage in any out-of-class activity they wished that had some relevancy to their educational objectives. They literally jumped at the chance. As they told me later, both had a new sense of freedom.

I was curious to see how they would handle their new freedom. Note that I felt reasonably sure that they would respond well. These two girls were serious, conscientious, and hard working. Now they could take courses for credit or sit in as auditors. One girl took fewer courses but did an inordinate amount of reading in her major field, French literature. Whatever French plays were given in the New York-New Jersey metropolitan area, she attended, reading the plays beforehand. I made available to her as many records as I could of dramatic and poetry readings by outstanding French actors and actresses. My friend Edouard Morot-Sir, since retired but then head of French cultural activities in America, helped me in getting some outstanding collections. We did the same thing with French songs and, within a short time, the young lady knew a repertory of over a hundred songs. I happened to have a sizable collection of music myself. Fortunately, the young lady played the piano and liked to sing. But a problem developed. Many old French songs, including some of the seemingly innocent children's songs, have a final stanza that is risqué. What to do. I told her, "You're a grown-up girl. You might just as well learn the facts of life. Besides, some day you'll be in France and your fellow students will be singing these songs and you might just as well learn to take things in stride." How prudish this all sounds in the light of present-day permissiveness and sexual frankness! She went to any French lecture announced in the French newspaper and attended every French movie shown within traveling distance.

For her reading, I encouraged her to build up her own library. I started her off by giving her about twenty of my books. As she read she underlined words she didn't know and, as she looked them up, jotted the meaning in the margin. At the end of books there are usually a few blank pages. She kept a double column of words, French on the left, English on the right, so that she could test herself on her rapidly increasing vocabulary. I would test her from time to time, a process she enjoyed rather than feared simply because she had faith in her progress and knew that inevitably I would be congratulating her. A whole new world was opening up for her and she was enjoying it immensely. She didn't neglect non-French fields, but these were on a less intensive level. Here she was motivated by the fact that if she went on to graduate work, balancing subjects would be expected of her.

Eventually she went on to study in Paris. I encouraged her to go to the Comédie française or to the Odéon every evening, if possible, but to be sure to read the play beforehand if she hadn't already read it. She averaged about five plays a week. She took courses at the Sorbonne, but in addition she literally devoured the museums and made sure she took advantage of every gallery talk available. She made friends easily and lost no opportunity to get into spirited French conversations whenever possible. Needless to say, in time she was ready to become a French teacher and a very good one at that. I wish I could say that she stayed long in that profession. She soon got married and acquired a new one—housewife. Now she has fun teaching French songs to her children, all the stanzas except the last one.

The other girl was interested in business. She just gobbled up business subjects and took more courses than required. I pointed out to her the necessity for general cultural subjects and she readily assented to the idea, although I made no demands upon her. As a matter of fact, she saw many relationships between English and social science and psychology subjects and her own preferred areas of study. She also understood the value of a good general education in the business field. But the very fact that she made her own choices motivated her work and, needless to say, in spite of her heavy program she made top grades in everything she took. She was a harsh judge of teachers. If she felt the teacher was good she gave him her all. But if she felt that the instructor didn't know his subject, or was soldiering on the job, she was outspoken in her thoughts and quickly dropped the person. Was she too severe in her appraisals? After thinking back over the haze of years, I must say she was right most of the time. I remember her vehement dislike for one instructor in particular. He was a dazzling character, full of good stories and an immediate success with the students. He gave the impression of being a world traveler, very sophisticated, and very much au courant on almost anything— a real leader of men. After a week or so in his class, our young lady announced that she was dropping his class. She said simply that this man had a glittering front but was a shallow person. Sure enough, by the end of the semester, the students became tired of his bonhomie and realized they were learning nothing. There's nothing like a female with insight!

Accounting became her passion, and, jointly with that—taxes. She had a mind that popped with facts and figures and ideas. She took a part-time job with an accounting firm in spite of her heavy load and immediately became their favorite. She was the sort of person who read the fine print and read it quickly, while others just let it go for some future occasion. It would have been a disgrace to try to hold back this girl to the plodding tempo of the average college class. And yet, this is exactly what we do with probably one-third of the students in any class.

Today we are attuned to any sort of new procedure that liberates the student from the shackles of classroom discipline. Until a few years ago this wasn't so, and it still isn't so with the great majority of the 2500 colleges in America. State and accrediting agency requirements tend to formalize curricular requirements, ever afraid of lowering standards. Faculty, curiously, in most cases tend in the same direction, usually in order to preserve the status quo or vested interests. The sensible answer to this problem is to allow flexibility with two provisos: one, that the mentor do a conscientious job of guiding his students, and two, that only those students who are mature enough and motivated enough be under this plan. This is exactly what we did with our Honors College, which I established just before I retired from the presidency. I must report that, while the faculty mentors in this unit did an outstanding job of guiding students, many faculty members and some chairmen did not make the task easy. I remember vividly how some years ago I was very close to acquiring a small campus in France. The people who were against the idea were the French professors who were teaching the advanced courses and were afraid that if the few French majors went to study in France instead of taking their courses, their jobs would vanish. The spectre of vested interest is much more serious than would be imagined. In large, rich institutions, the problem of a freer curriculum is solved usually by expanding the number of professors in proportion to the number of students. In less affluent or small institutions, the fear of vanishing jobs haunts faculty members and hence their inclination to conservatism. In passing, I must comment that in my long years of experience in education, I have never seen any faculty member dropped because his job had vanished. Obsolete courses are kept in the catalogue and given for few students, but the sanctity of existing jobs is always preserved. About the only flexibility that exists is not to fill a place when the professor retires, but even here there is often a younger replacement who has spent long years to fill that very job and will exercise every possible pressure to capture the plum.

So far, there has been such a tremendous expansion of higher education that there has been room for all; furthermore, the older private institutions have had enough money rolling in from donors to make up for academic featherbedding. But now the sources of money are drying up. For State sources, the era of retrenchment has set in. As for private institutions, we have slowly priced ourselves out of the market, and have allowed our costs per student to skyrocket, in many cases needlessly. But, getting back to the student, three things will happen: one, he will want greater freedom of choice; two, he will resist any appreciable rise in tuition; and three, he will demand greater performance by his professors.

In the early years of the college, I used to have faculty members visit students at home, for I wanted them to have some concept of neighborhoods and sociological patterns. I wanted to encourage the faculty to be not merely purveyors of information, which in most cases can be found in printed form, but to be interested in their students as human beings. I made it a practice never to ask the faculty to do something I wasn't ready to do myself and so, of a Sunday morning, Sally and I would do our share of home visiting.

I remember visiting one home in Ridgewood. As the door opened, the mother burst into tears. Sally and I were somewhat taken aback by the unusual reception.

"My son is doing badly?"

"Not that I know of," I replied.

Then she told me the story. That month her husband had run off with his blond secretary. It had been a traumatic experience for her and for her son. She was sure that he had fallen apart and that this was the reason for my visit. I assured her that the visit was routine, but that I would look into the matter the next day, which I did. Sure enough, the boy was having trouble, but when the faculty learned the cause, they gave him the helping hand he needed during the family crisis.

One faculty member, Professor Bogdan Raditsa, arrived at a student's home and found about forty guests there. He apologized for breaking into what seemed to be a family party.

"Oh, no," the father replied. "Our son is the first member of our family to go to college. The visit of a professor is a great honor for our family. We all wanted to be here to greet you."

When the professor told me this story, there were tears in his eyes. How could he help being as good a teacher as he could to the young people in his charge?

At the end of a visiting day, we would all gather in the castle and, over cocktails and supper, would recount with joy and vivaciousness the experiences of the day. We knew our students, we were aware of their problems, and our one desire was to help them as much as possible. Has the day for such close contact with students passed? I don't think so. I think it is still possible for private institutions to give this kind of service and even for large institutions to do so, provided that they break up the large masses of students into smaller groups of, let us say, not more than 400, as we tried to show the way in establishing Edward Williams College.

17
Undersea Capers

Sometimes opening a campus overseas can lead to some highly diverting experiences. We were to have the dedication of our marine biology laboratory at St. Croix, Virgin Island. The idea had grown out of a biology student-motivated series of trips to the island. On their own, but led by a creative teacher, Antoinette Anastasia, they had decided to spend ten days enriching their biology studies by looking for specimens in the beautiful Buck Island area off the northeast coast of St. Croix. Senator Fairleigh Dickinson happened to hear of the project and became intrigued with the idea of having a permanent marine biology station. One idea led to another and soon we were drawing up plans for just such an establishment, which would not only give a broader outlook to the students but also could lead to the establishment of an oceanology curriculum at the University. At any rate, Senator Dickinson gave eight acres of the land he owned in the northeastern part of the island and pledged the money for the building. Faculty members began to work on the course of study. This is how a university develops. Note that the initial impetus came from a group of enthusiastic students and a stimulating teacher. When the project was a reality, it was decided to have a dedication ceremony. Rather than have a dry, pompous function, I thought it would make more sense to have a conference on oceanology. By this time I had become fascinated but appalled by the unintelligent use of the sea and felt that a lot of academic people would share my feelings. As I was to say later in my brief remarks at the dedication: "If the human race were wiser, we would stop the crazy race for the moon and concentrate on the intelligent use of the sea. But the human race is not wise and so we expend our greatest energy on space, and slower pace to the sea, and probably somewhere in the seventies or eighties millions of people will suddenly die from starvation and then, and only then, will we awaken to the realization that the sea will help us to prevent future self-annihilation of the human race.

"It is almost with a sense of impending holocaust that faces the world, that I dedicate the marine biology laboratory of Fairleigh Dickinson University as a step toward the future, for no university worthy of its name can afford not to concern itself with man's last effort for survival."

One of the trustees, Leonard Dreyfuss, gave me a grant to run the conference and I was able to get some of the outstanding leaders in oceanography to lead the discussion. Rather than have a cornerstone laying, I conceived the idea of laying a statue at the bottom of the sea not far from the shore.

I have never been one to wait long years to lay out plans. I have found that if you make plans too far in the future, everybody waits until the last month anyway, and then there is a frenzied attempt to reach the goal. The only thing that happens when you wait too long is that you give yourself a chance to forget what you learn and lose the momentum of action. So, in this case, we gave ourselves two and one-half months for a conference June 20 to 22, 1967. But almost two months passed by and I hadn't found the statue I wanted. I did have a standby in case I didn't find anything, but I would have considered it a flat failure if I was forced to use it.

As chance would have it, we attended an outdoor luncheon given by a member of the Board of Fellows, Saul Rosen, in his lovely garden in Paterson. Saul is one of the great art collectors in New Jersey and he has many outstanding sculptures in his garden. On our way home, Sally said, "You know, that sculpture by Chaim Gross wouldn't be bad for St. Croix." Through the years, our minds have had a sort of telepathic connection. As a matter of fact, once, having heard my friends Dr. and Mrs. David de Sola Pool give a duologue in which husband and wife complemented each other by picking up the thread of discussion almost imperceptibly, I decided to try the same thing on Sally. In the middle of a speech I was giving at a local high school, I stopped and asked Sally to continue. Sally was startled but since she knows everything I know, she collected herself and spoke on for ten minutes or so, when I stepped up to the lectern and finished the talk I had started. In this case, hardly had she finished the sentence when I said, "That's just the statue and that's what we're going to use at the dedication."

When we got home I called Saul and asked him whether he would have any objection if we had a copy made. He didn't. I called the sculptor whom I knew and asked him whether we could have a copy molded in one week. He said, "That sounds crazy but why don't you call the owner of the foundry and tell him he has my permission to turn out a casting." I called the foundry owner and he, too, thought my request was insane, but after some urging on my part, he agreed to start the process on the morrow. In one hour we had made a major decision and worked out the mechanics. In the meantime, my mind was working on the type of base we would need.

The next day I was playing golf with Steve Van Zandt, a contractor. I asked him whether he could help me with the base, knowing that he would be only too glad to cooperate. By the time we were through we had figured out the approximate dimensions. He went to look at the original casting in Saul Rosen's garden and proceeded to cast the block, which had to be heavy enough to sink to the bottom of the sea—about eighteen feet down at the point we had in mind. Simultaneously I had a bronze plate made with the legend:

Man returns to the sea for knowledge and abundance.

The next few steps were complicated. The plate had to get to Steve so that he could bolt it on the concrete. The base from Steve's shops and the statue from the foundry had to get to the air depot at the same time. Someone had to supervise the shipping to St. Croix. There are certain times in life when it's more trouble to delegate than to do it yourself. This was one of those times. I called my friend Ralph Paiewonsky, Governor of the Virgin Islands. Could he help me on the other end? "Well, Peter, I want to see your conference succeed so I'll have a truck pick up the statue and the base. We'll keep them until you're ready for the ceremony. A derrick boat will pick up both pieces and drop them wherever you wish." This was a Governor after my own heart.

Now I had to get an undersea diver to bolt the statue to the block. I found one in Morris Lee, who got into the spirit of the occasion. The day before, we had a dress rehearsal. Miraculously the derrick boat appeared with the base and the statue. The press and television people were there to take advance photos. Steve Van Zandt had made a special trip to supervise the mechanics. He wasn't going to miss the fun. The base was lowered without any trouble. The statue came next. Gingerly it was placed on the base. Lee immediately lowered himself and bolted the statue on the base. At this point the press wanted pictures. Would I dive in past the statue? Who, me? I hadn't tried to tread bottom for forty years! I jumped in with the most awkward plop; the photographers realized I was not their man. I came up and turned to one of the trustees, Henry Becton, to take my place, which he did with infinitely more grace and precision. Weeks later the photograph was to appear in the Bangor News and Henry was the recipient of many congratulatory remarks on his professional aplomb. Lee then unbolted the statue and it was hoisted up.

The day of the dedication came. All of us had gathered at the St. Croix Yacht Club. A boatload of board members and a few dignitaries and the press people went out to the spot inside the reef, hoping that the derrick boat would again come along. It did, thanks to Governor Ralph Paiewonsky. The base was lowered first, then, slowly, the statue, wrapped to protect it, was lowered. Lee began to bolt it in. But before lowering it, we had brought along a bottle of Crucian rum which we felt would be the proper liquor for the christening. Then one of the trustees, Martin Weiner, said "Why waste a whole bottle? A few drops will do for the christening. Let's drink the rest of it." The idea was wildly applauded. Everybody felt in a devil-may-care mood. It was an utterly crazy, informal, and ecstatic dedication and celebration. If ever a group of men were thoroughly relaxed, thoroughly enjoying themselves, this was it. Singing and laughing, we chugged our way back to the wharf, where the rest of the crowd was assembled. Then we simmered down and a thoroughly respectable and sedate ceremony took place, with a few speeches. Dick Dickinson hosted a generous cocktail party at the Yacht Club. It was an auspicious beginning.

Months later, Sally sent a photograph of the undersea statue to Chaim Gross. He came to see us distraught. "The statue is on backwards!" We promised him we would have the error corrected. In due time, Lee Morris was able to dive undersea, unscrew the statue, turn it around, and place it correctly.

My friend Dorothy Gordon, moderator of the N.B.C. Youth Forum, made two color television programs during the conference. In the short weeks of preparation we probably had more experts of one kind or another than any other similar conference I have ever seen. Al Perlmutter, the television producer, put his heart and soul into the programs because he was a skindiving enthusiast. He and Dorothy made a wonderful team. The two programs they shot were shown twice on N.B.C. TV stations and broadcast on N.B.C. radio in the United States and also in Canada and then repeated. Campbell Norsgaard and Lou Ciampi, one of the world's best underwater photographers, as a favor to me knocked out a movie short on a budget no larger than a home movie project. This was shown hundreds of times in dozens of local movie houses throughout New Jersey and at many community meetings. We must have got a half million dollars worth of coverage without spending more than the few thousand dollars that Leonard Dreyfuss contributed. But what was important was that hundreds of high school students were brought into the forums Dorothy arranged and they had a chance to meet some of the greatest oceanographic experts of the world. Dorothy was thoroughly exuberant. She was over eighty at the time and yet she went scuba diving and came out on the beach brimming with smiles as any teenager would. For months at N.B.C., from the president down, they would greet her, "And how is our water nymph doing today?"

Official representatives from fifty-one colleges and twenty-seven professional and learned societies, government agencies, and industries attended. Members of our Board of Trustees and of our Board of Fellows also attended.

Robert Abel, head of the office of Sea Grant Programs of the National Science Foundation, came from Washington to bring us up to date on the government's participation in oceanology over the years and also to integrate academic activities, industrial needs, and governmental interests. Incidentally, we used the broader word "oceanology" rather than the more confining term "oceanography." Dr. Milner B. Schaefer, Director of the Institute of Marine Resources of the University of California at San Diego, took up the problems of food from the sea and the economic and political implications. He also delved into the field of mariculture and in sessions with the adults and later with young people he had all of them fascinated with the problem. Dr. John H. Heller, president of the New England Institute Graduate School, took up different facets of the same general problems. He stressed the fact that many people are not used to fish food and do not like anything made from fish. Indeed, we had had fish flour sent up from the Department of Agriculture and the chef at the Grape Tree Inn made it into bread and cookies for the closing dinner. But we didn't tell the guests until they had eaten both, otherwise they would have either spurned the items or imagined a fishy taste. He had us fascinated with his stories of his experiments with sharks. The subject of DDT in fish was explored. Alas, it wasn't until many months later that it began to dawn upon government agencies that the poisoning was much more serious than anyone had thought. Mr. Robert D. Gerard of the Lamont Geological Observatory at Columbia University took up the experiments using condensers in getting nutrient-rich water from the ocean to meet the drinking and industrial needs of Caribbean islands. They had envisaged such an apparatus on a floating platform or ship. The costs of processing water were then taken up and these had immediate relevancy to the island where we were meeting. And then we went into the fascinating problem of ocean engineering. I don't think anyone could have touched on this more dramatically than Willard Bascom, president of Ocean Science and Engineering. His discussions of operations from Vietnam to South Africa had us all enthralled, especially when he spoke of mining diamonds offshore in Southwest Africa. And when he spoke of the tighter control the Russians had over their men as compared with the shorter and more permissive tours of duty of American engineers, one sensed the advantages the Russians enjoyed on the international oceanographic scene.

Dr. Arthur E. Maxwell, Associate Director of the Woods Hole Oceanographic Institution, introduced the whole subject of the international aspects of oceanography and, indeed, what he took up was the basis of the discussions held under the aegis of the United Nations. Shortly thereafter I was to make the acquaintance of Dr. Arvid Pardo, the Ambassador from Malta to the United Nations, who was to spearhead this very question at that body. Dr. Maxwell's presentation was another example of the relevancy of the conference to everything that took place in the world. Should the national sea limits extend three miles or 200 miles as in Peru? No international body had more knowledgeable experts than the conference that was held and every minute was breathtaking. This was especially brought out by Dr. William B. McLean, Technical Director of the United States Naval Ordnance Test Station, who introduced the whole area of travel and living undersea. When he got through we all felt that we had been hearing about another world.

There wasn't a dull moment during the three days of the meeting. Each one of the speakers was dynamic and forceful. The questions and answers were especially fruitful. When we were through, members of the Board of Trustees and of the Board of Fellows who were attending met and tried to work out the implications of the conference for the university. Later, a volume was published on the papers presented and the discussions: *Conference on Oceanology.* It is now in the libraries of over one thousand colleges and universities. It was the most fruitful conference any of us had ever experienced in educational circles.

The new $9,000,000 School of Dentistry in Hackensack (Architect's drawing).

The Florham Mansion on the estate created by Florence Vanderbilt Twombly at the turn of the century.

President Yameogo of Upper Volta, West Africa, being made an Indian Chief. Left to right: Chief Thunderbird of the Shinnecock Tribe, President Yameogo, Dr. Sammartino.

St. Croix, Virgin Islands. Instead of having a ground-breaking ceremony for the dedication of the Marine Biology laboratory at St. Croix, we placed a sculpture by Chaim Gross at the bottom of the sea with a dedicatory plaque that can be read from the surface of the water. It was here that the newsmen asked Dr. Sammartino to perform for them.

Sculpture by Chaim Gross, later bolted to a concrete slab and placed on the ocean floor. Governor Ralph Paiewonsky and members of the Board of Trustees.

Wroxton, England. Wroxton Abbey, the ancestral home of Lord North.

Wroxton, England. Dean Loyd Haberly lecturing to students in the Regents Room, the most beautiful classroom in the world. On the walls are beautiful Scalamandré silk and linen fabrics re-created from the original gold brocades. On the mantel are antique vases donated by Mr. and Mrs. Thomas Kelly.

Wroxton, England. Some of the presidents from around the world during one of the sessions held in the Great Hall of Wroxton Abbey.

Wroxton, England. The academic procession as it leaves the Abbey and passes the First Army Band. Leading is Dean Loyd Haberly, carrying the English Sheffield silver mace. Dr. Heinz Mackensen, Chairman of the University Council, carries the Wroxton banner. Then follows the Lord Bishop of Oxford. Dr. Sammartino accompanies Lord Harlech.

Edward Williams College, which Dr. Sammartino opened in September 1964. Sculpture of Ulysses by Mestrovic can be seen in the foreground.

In 1968 Dr. Sammartino organized the second triennial meeting of the International Association of University Presidents in Seoul, Korea, and led a group of college presidents to visit universities in nearby countries. Here, Dr. Sammartino and Dr. Adolf Robison are having a chat with Generalissimo Chiang Kai-hek at the palace in Taipei.

The Fairleigh Dickinson monument in the Tiger Balm Garden at Singapore, erected by Dr. Aw Cheng Chye.

Dr. and Mrs. Sammartino at the establishment of the Peter Sammartino College of Education. Dean Donald Herdman is seated at Dr. Sammartino's right.

Dr. Sammartino enjoys taking elementary school children through the New York Cultural Center. In the background: Mrs. Sammartino speaking to Mrs. Smadbeck, president of the Heckscher Foundation, which helped in setting up the project.

The sculpture Peace Screen by one of the alumni of Fairleigh Dickinson University, Paul von Ringelheim, placed at the United States Pavilion at the 1964 World's Fair. Dr. and Mrs. Sammartino with President and Mrs. Young Seek Choue of Kyung Hee University in Korea.

Dr. Sammartino, accompanied by Dean Marinus C. Galanti, Dean Clair W. Black, and Mrs. Sammartino, opening Fairleigh Dickinson University Day at the 1964 World's Fair in New York. The University Bagpipe Band leads the parade.

18

Wroxton

Many of us have dreamt of being naked in a crowd of people but I shall never forget when this actually happened to me in our Wroxton Abbey in England. It was right after the dedication of our English campus. It had been a hot day and I had perspired profusely. Sally and I had a room on the second floor in the far end of the Abbey, but our sumptuous bath was way over on the other side of the mansion and intervening was a secret passage leading to a long hallway to the bathroom. There is a door on the first floor, supposedly kept locked, but the door looks like part of the wall and no one, unless told, would know that it was a secret entry to the apartment above. I undressed and went to take my bath. On my way out, I heard a babble of voices. Here I was trapped in the bathroom with no clothes. As a matter of fact some of the guests were even tugging at the bathroom door. It was like one of those Freudian dreams when one feels entirely naked at a cocktail party. Finally, I did the only thing I could do. I screamed at the top of my voice. "I've taken a bath and I'd like to get through to my room. Will you please clear the way or at least close your eyes." They scampered away and I scampered through, my pudor intact but my dignity slightly shaken.

But let me go back and tell the story of the establishment of our campus in England. Ed Williams, the Chairman of our Board, had been visiting his married daughter, who was living in England. In the course of some sightseeing he had seen Wroxton Abbey, about twenty-six miles from Oxford, and had been impressed by it. The Abbey is situated in the delightful village of Wroxton, most of the houses of which are thatched roofed and have lovely flower gardens. It is on the Stratford road in the enchanting Cotswolds, which many consider the most historic and the most beautiful part of England.

The original priory had been built in the thirteenth century, under a charter by King John. In the sixteenth century, Henry VIII, as part of the dissolution of all monasteries, had the walls of the church pulled down. The Wroxton priory passed by purchase to Sir Thomas Pope in 1537, a local person who had risen high in the King's favor. When Sir Thomas founded Trinity College in Oxford, he endowed it with the Wroxton property of some 3000 acres. To this day, Trinity College collects the income from the property and, of course, it was this College that sold us the property in 1964. In 1618 a descendant built the Jacobean mansion which is now our Wroxton College. Various additions have been made through the centuries, but, by the beginning of the twentieth century, photographs show the rooms to be cluttered in the worst of Victorian style. All of this had been cleared out by the time Lady Pearson rented the building from Trinity College. When we got it, all I can say is that it was a mess. The pews in the small chapel were still in place; most of the woodwork in the great hall, in the large music room, and in a few of the parlors was intact, thank God. Literally, we didn't know where to start, but start we did. If I had to do it over again, I just wouldn't.

Later on Ed Williams's daughter was visiting him in the United States and brought with her a copy of the *London Times* containing the announcement that the Abbey would be put up for auction by Trinity College of Oxford, which owned it. It was during the summer of 1963 and I was at East Sullivan, Maine. Ed called me excitedly to tell me about Wroxton and of the need for quick action in case we were interested.

Curiously, on the evening that Ed called, among our dinner guests were Mr. and Mrs. William B. Schieffelin, who had been with us when Ed had called us in 1957 to tell us about Florham. I turned to Bill and said, "For Heaven's sakes, Bill, it seems that every time Annette and you visit us, we are about to acquire a new campus."

I proposed that Dean Loyd Haberly be sent immediately to Wroxton and give us a report by telephone. Loyd had studied at Trinity and, as a matter of fact, the president of Trinity, Arthur L. P. Norrington, had been his classmate. I called Loyd, and within a few hours he was on his way to Wroxton. He telephoned me at East Sullivan thirty hours later to report that the Abbey was indeed a gem. A Lady Pearson had been leasing it from Trinity and had a half-dozen assorted tenants within the Abbey. As a historic monument, the building was open to the public for 2 shillings sixpence, and Lady Pearson also ran a little tearoom in the dining-room of the Abbey. She had furnished it with sundry pieces of furniture and altogether it added up to a modest if precarious living for her. Many similar buildings in England are so operated.

Wroxton Abbey had been the ancestral home of Lord North, that much-maligned prime minister of George III during our own Revolution. In Adlai Stevenson's speech written for the occasion but read at the dedication by me because in the meantime Stevenson had died suddenly, he says:

> If the shade of Lord North haunts Wroxton Abbey, I would like to think he is chuckling a little at this latest audacity of the upstart colonists who dared to buy a property from Oxford and use it for the purposes of American education.

As I write this, C.B.S. is featuring a special on Lord North being interviewed on the American Revolution by the commentator Eric Sevareid. Peter Ustinov plays a delightful Lord North, sitting in the Great Hall of the Abbey, surrounded by the furniture and my suit of armor sent from Bergen County, New Jersey, when we took over.

Loyd's report was that since the building was already being used by tenants, it should be usable for students and that we should be able to buy most of the furniture from Lady Pearson. With the expenditure of some money for the students' rooms and other classroom appurtenances, we should be able to operate. It didn't quite happen that way, and before we were through the expenses had soared to ten times the amount we had considered as a result of the telephone calls. There were good reasons for this.

One of the virtues of the English people is to "make do," much in the same tradition that I had observed in New England. A chair is covered and recovered, and this can go on forever. A "fridge" or an electric heater will be used to the very moment when it collapses from sheer circulatory ailment. In the process things may get to look down at the heels and the aesthetic effects leave something to be desired. You can't quite operate a college in the same fashion, although I have seen many schools in England that had a dismal, down-at-the-heels look. I had to make a decision, and it was that Wroxton Abbey would look like a model school aesthetically, even while preserving the seventeenth-century atmosphere. Furthermore, since this was a historic building, we had to check at every stage with the British Historical Commission.

Although I have the dubious distinction of having turned more old mansions into colleges than anyone in the history of higher education in America, I had no experience in restoring an old English Abbey. I have always found that common sense and working away, plus money, can solve most problems. Many of the old castles in Europe have been honeycombed with special apartments, some of which wind in most peculiar fashion. Kitchenettes and toilets are squeezed into the most unlikely places. But let us go back to the actual acquisition of the Abbey.

After Loyd had made his inspection, he notified Dr. Norrington that Fairleigh Dickinson University was interested in buying the property. This pleased the people at Trinity, since the only other source of interest they had had was from a commercial organization which would have changed the whole aspect of the Abbey. We made an offer of £30,000, which was agreeable to Trinity, and they immediately called off the auction with no complaints from anybody. For approximately $85,000 we now had an honest-to-goodness old English Abbey with fifty-seven acres of grounds, including two small lakes. In addition, there was the old carriage house, a building of no mean proportions, plus a delightful caretaker's cottage at the entrance of the estate. All very nice, except that practically everything was in disrepair. The more one poked around, the more things were found that had to be fixed. Little by little, it dawned upon us that this was going to be a bigger job than we had anticipated.

I wanted to gain some perspective and I thought it might be of value to see what John Paul Getty had done with the estate he had bought at Surrey. He graciously invited us to visit his place. Having unlimited funds, Mr. Getty could of course do anything he chose. Everything was in exquisite taste and I found that he understood more about styles, periods, and details of architecture than people generally realize. Every detail of interior design or decoration had to be exact. Nothing was second rate. The paintings were by old masters, the tapestries genuine antiques. I found only one drawback, and this was to lead to a major decision regarding Wroxton. Every room, with the exception of the small informal living-room, was beastly cold. Since I couldn't keep my hat on, my head was uncomfortably cold and my teeth were actually chattering. I made up my mind that Wroxton would be the warmest school in England. Mr. Getty, of course, kept the heat down because of the astronomical value of his paintings, which might deteriorate at high temperatures. Curiously, instead of being discouraged when I got through with the visit, I felt encouraged. While I couldn't do things in the grand manner of Mr. Getty, I had at least a standard to go by.

Back in New Jersey, over the years, we had accumulated many paintings, sculpture, pieces of antique furiture, and objets d'art. I immediately began to make mental notes of the many pieces I could send over to Wroxton. Usually antiques cross over to America. This was to be a process in reverse. In the end, we were to send over about $300,000 worth of works of art. It was a tremendous problem in logistics, one which I could not have accomplished without the invaluable help of Al Herald of the university staff in Rutherford. There was a handsome Gothic table I had bought from the estate of Dr. Gregory in Manhattan and had given to the University. It had been a beautiful addition to the castle in Rutherford. Now it is equally beautiful in Wroxton. Sally's father had given us another extraordinary Renaissance table, which he had had in his office. That too went overseas.

One room in the Abbey was the bedroom occupied by Charles I and James I. It had been used by one of the tenants and was in a deplorable condition. We restored it to its original design. For furniture, we sent over an antique bed we had bought from the Twombly estate. We also sent over a traveling desk used by Napoleon, given to us by Tom and Sarah Kelly, and this leads us to another story.

One evening we were having dinner with the Kellys in their charming town house in New York. We spoke of Wroxton and, since Sarah is one of America's great interior decorators, she immediately became interested. Tom is primarily a lighting expert, but he is also involved in decorating and design. We had them in hysterics with all our problems in Wroxton. The Kellys had had a magnificent chateau in France, which they had partially dismantled just before the Germans overran France. Halfway through our discourse, Sarah said, "Peter, I have some things that I brought home from our chateau. Perhaps you could use some of the antiques." We immediately descended to the lower floor and I literally leaped with joy. There was a gilded lantern three feet in diameter that was just the thing for the grand staircase. A precious sixteenth-century baptismal font was just right for our small chapel. Two Renaissance vases were perfect for the enormous fireplace of the Regent's Room. Even Mr. Getty couldn't have much better. Other antiques followed, one of which was the historic desk mentioned above.

Back to the King's Room. We soon had a few more pieces and these, plus a silver service donated by my friend Burl Ives, make a presentable King's Room—even though it lacks one prime necessity. It has no connecting bathroom. Visitors incorrectly assume that the contents of the room are originals. They are somebody's original, but not King James's or King Charles's. We went through a similar process with the nearby Queen's Room. In between is a small room, furnished in contemporary style, for a single student. It may sound strange, but somehow it all fits together.

As I have stated, I decided that this was one school where students would not shiver to death. Central heating of sorts already existed but, as anyone who has spent some time in England knows, such systems work very spasmodically and even at best offer little competition to the inevitable drafts in large homes. The heat is barely enough to keep the pipes from cracking. Such was the case at Wroxton. The biggest problem we encountered was that of hiding the radiators in order not to spoil the ancient woodwork. This we did, with the invaluable help of our architect, Mr. Cunningham.

Reluctantly we decided to close off all fireplaces. It was just too much of a fire risk. We put in electric heaters instead, and these are operated by inserting a shilling piece. Knowing the wastefulness of students or, for that matter, of any group of people living either in hotels or dormitories, we took the precaution of putting in shilling meters—one hour of current for 14¢.

Through the courtesy of Mr. Ricky Davis, then assistant bursar at New College at Oxford, we were in touch with Minty's in Oxford and were able to solve the problem of furnishing the student rooms and the library. We made decisions quickly and, while the tempo of such dealings with English firms is much slower and is tied up with a great deal of protocol, they soon got into the spirit of urgency and cooperated efficiently. From a nearby chair company, I ordered specially made tablet-arm chairs that fitted in with the somber and elegant music room. I took a version of a captain's chair and had a tablet arm added, just wide enough to take a pad or book. I was amused to learn later that the company had displayed this chair at a commercial exhibit, whereupon a number of tavern owners ordered the chair as just the thing for their establishments. One never knows how education affects the rest of the world.

The North family was Catholic. In the period when Catholic services were banned, there was a hidden room for the priest where he celebrated Mass. I had a small altar set up and the carpenter made a small crucifix for it, the idea being to give some concept of the hidden sanctuary. Historical signs had to be prepared for the many interesting rooms.

As I said before, Lady Pearson had divided the abbey into a number of apartments. Most of the tenants moved out without any trouble when we took over. But in one corner of the castle lived an "artist" and his wife, and they turned out to be a major problem. Somehow they had appropriated a series of rooms and had proceeded to fill them with junk and newspapers. Finally Mr. Davis, by paying for their moving expenses and giving them a payment to boot, was able to get them to move, but it took months. The stench and the dirt we had to contend with were unbelievable. Today those very rooms are light and airy rooms painted in gay colors and furnished in contemporary style.

My one fear was of fire, so we took every possible precaution. As I said, all fireplaces were closed up. A complete fire alarm system was installed. No smoking was allowed in bedrooms and the reason for this seemingly strict rule is explained to students. Fire extinguishers are everywhere. Fire drills are held regularly, from the very beginning of each session. I have always insisted that drills be held in the dead of night in spite of the discomfort they cause. During the dedication, NBC had sent over a crew to tape two programs. One of the secretaries almost set the Abbey on fire before we even started. She was smoking in bed in a midday siesta, fell asleep, and, before she knew it, her bed had caught fire. Luckily, the blaze was extinguished almost immediately. But the scare taught us a good lesson.

I don't think that Sally and I ever worked harder than we did in getting Wroxton into shape. While we had lots of different workers and artisans, each one belonged to a separate union and of course was not allowed to do any thing outside his special field. The result was that if any furniture had to be moved, Sally and I had to do it. Added to all this was the fact that the Abbey was like a great big cold cave. The chilling English weather, abetted by even chillier currents of air that swept up from the basement, penetrated to our very bones. I partially solved the problem by having three electric heaters in the minstrel gallery, which was chosen as the permanent office. We would get as warm as possible, then rush out to do as much work as possible until it became too cold, and then rush back to get warm again. They were hard days, but exhilarating ones too. Who would ever have thought that we would be putting an English Abbey into shape? We were filled with a sense of English history and almost a childish glee in feeling that as Americans were were reestablishing the grandeur of the castle.

At night we would go back to the Whately Hall Inn in nearby Banbury and soak deliciously, ever so deliciously, in the hot steaming water of the bath. Afterwards followed the dry martinis at the bar, made to my specifications. This is not so pompous as it sounds. In most English hotels, cocktails are minuscule. If you want a regular-size drink, you must order a double. We usually have two one-and-a-half-ounce drinks every evening. To solve the problem at the Whately, I had the young lady at the bar fill the shaker with ice, add five jiggers of gin (each of one ounce), and add another jigger of French vermouth. This was a minor revolution in the quiet decorum of the downstairs bar and always excited the interest of the other customers at the bar. Now friends who happen to visit the hotel can ask for a dry martini as I used to have them made. Otherwise, one is liable to get a glass of simple French vermouth.

After dinner, if we didn't have guests, we would go back to work at the Abbey, getting a few pesky chores out of the way. To be working all alone in a spooky old castle at ten in the evening isn't what most people would call fun. But it had a strange sort of exhilaration for us, and when we pulled into bed at night we felt that we had done two day's work in one.

As the dedication day approached, there was one small room that could not be finished in time because of the unexpected amount of dry rot that had been discovered. Sally had a bright idea, as she usually does. Why not leave the wall exposed so that visitors could see how dry rot appears, and have a neat little sign explaining it? We did exactly that, and the room drew the interest of all the visitors.

The problem of dry rot was particularly acute at Wroxton. We would find, for instance, an indication of dry rot in a beam in the basement. Before we knew it, we had traced its course upwards to the second floor. What started as a simple job now had turned out to be one of long duration and considerable expense. This was by far the most annoying problem at the Abbey.

Many small decisions had to be made on the spot. For instance, in a stained-glass section of a window on the first floor, one of the panels had been broken. I had a new glass panel with the new Wroxton coat-of-arms made up and inserted in its place.

The Regent's Room deserves special mention. This was made for the Prince Regent's visit in 1805. It is a spacious room in gold and off white. I call it the most beautiful classroom in the world. We were able to have it restored by true artisans with real gold leaf. Over the fireplace was placed a beautiful painting à la Fragonard; at the sides were the two beautiful vases from the Kellys. But it was from my good friend Franco Scalamandré that I received the most desperately needed help. I appealed to him to help me with the old silk golden wall coverings and the matching drapes. Franco went all the way. He had the historic wall coverings and drapes recreated and woven and presented them as a gift to the University. Conservatively, it would have cost us $80,000 if we had had to pay for these items. The chairs I had made in gold and red.

The piano story should be told at this point. We had to go back to London and stopped at the Hilton. While we were having cocktails in the bar, I noticed an Indian sitting all alone. Being of a gregarious nature, I invited him to join us. He gladly assented, but added that he was expecting a doctor friend soon. I proposed that he join us too. The doctor arrived in a few minutes and naturally our conversation veered to Wroxton. During the talking I happened to say that I wished I could find a piano for the abbey. "I've got just the place for you," said the doctor. "It's a small shop in the East end of London and the man is dying to get rid of this big piano which is crowding his small quarters." The next morning we were at the shop. The piano in question was a beautiful Bechstein grand with superb tone.

"How much?"

"Two hundred and fifty pounds."

"Refinished?"

"Oh, completely refinished." We bought it on the spot. On the way out, I saw an attractive grand piano case in gold and white—just the thing for our Regent's Room.

"What's that?"

"Oh, that's a prop that's often used in movies."

"Can you put in the piano works?"

"Of course."

"How much would it cost with the works?"

"Oh, about one hundred and forty pounds."

"OK, we'll buy it."

Before I left I had also arranged for him to fix the small organ in our chapel for about fifteen pounds. All this had taken about fifteen minutes. The two pianos would probably have cost us $7,000 in the United States.

One of the first things we did was build up the library. Loyd Haberly gave the specifications for the books to Sir Basil Blackwell in Oxford, who set his staff to the task and both Loyd and I were pleased with the result. We had one important stroke of good luck while doing this. The library of C. S. Lewis came on the market. We were asked whether we were interested in buying it. We were, and we commissioned Blackwell to purchase it at a reasonable price. This was soon done and added immeasurably to the educational value of our library. In addition, Althea Herald ordered books in America and the faculty at Wroxton made up other lists. I also ordered a whole collection of interesting books about England so that students could read about almost any subject as they visited the different parts of the country. Last, I asked that a complete collection of English magazines and newspapers be subscribed to, in order to encourage wide familiarity with all phases of English life.

The educational phases were the easiest to work on, and the most pleasurable. For instance, we wanted to involve students deeply in the Stratford Theatre, since Wroxton is on the Stratford road and only 16 miles away from the supposed birthplace of Shakespeare. We were fortunate in making very close connections with many of the outstanding Shakespearian scholars and with persons who were producing the plays at Stratford. The students received full explanations of the plays before they saw them. Of course they read the plays in their entirety and then when they saw these plays beautifully presented on stage at Stratford, each visit was memorable one. The whole Stratford experience was not without some traumatic effects for me, however. I am one of those rare birds who do not believe that the true Shakespeare is the lout from Stratford. I am an Oxfordian; in other words, I believe that the true Shakespeare was probably Edward de Vere, the Seventeenth Earl of Oxford. Out of 2200 college presidents in the United States, I don't know of any one of them who is an "Oxfordian." It is somewhat ironical that the only president to have a campus in England near Stratford and to be dedicated to the task of preserving the sanctity of the Stratford Shakespeare should have been I. I took it with a sense of humor, however, and it never interfered with any of the lectures, with which I disagreed violently. I sense that Loyd Haberly was always somewhat fidgety lest I begin to argue with some of our visiting Shakespearian specialists.

We mapped out a whole series of trips to other parts of England so that the students would get orientation in the history, the geography, the archeology, and the industrial aspects of England. We also wanted them to have a thorough idea of English institutions. The visiting faculty members were outstanding. I polled the students on closed ballot votes a number of times through the years. Out of about twenty, there was only one who did not go over, and, of course, he was not invited back to lecture. I must give the main credit for the academic excellence of the institution at Wroxton to Loyd Haberly, who pursued his task with dedication, with fervor, and with patience.

The library, of course, was the heart of the Abbey, and rarely was it seen without a whole group of students reading away or working at their reports. It was a pleasant place to be in; it was inviting, and I am sure that the students had the same experience that I had— the spirit soared as one looked out the spacious window to the green meadows beyond. As a matter of fact, I purposely had a nearby farmer graze his sheep in our meadows so that they could be seen gamboling as we read in the library. The students were busy every minute of the day, and they loved it. How wonderful it would be if every student in every American college could have such an experience! And their own rooms were similarly conducive to study and writing. There was something in the Abbey that promoted an intellectual frame of mind without in any way interfering with the conviviality that should obtain in any academic setting. As a matter of fact, Sally and I took special pains with the student's lounge room. We sent over beautiful original paintings of value for these rooms. It was a living-room atmosphere conducive to sophisticated and mature conversation.

The outside visits were always very pleasant and there was always a gay camaraderie. Often picnic lunches were taken along and nothing could be gayer than a roadside picnic. In our dining-room we followed the English custom of having a high table for the faculty.

We were fortunate in having a diminutive chapel that was really a gem. As a matter of fact, the stained-glass windows are among the best in England. I added religious paintings by de Luca and, of course, the beautiful baptisimal font given by the Kellys. I shall never forget the first chapel meeting that was organized by the students themselves, which happened to take place on July 4th. It was nonsectarian in character and the hymns were of a national character. I believe I can say without contradiction that all of us, students and faculty, felt the spiritual unity regardless of individual religious differences. At various times I have seen students go into the chapel to pray. Another thing I did was to install a stereo outfit for the Great Hall so that the students could hear plays, or listen to English folk music or any other music they chose to. In order to encourage students to go to see the countryside, I put in twenty bicycles that could be rented at reasonable prices.

For the dedication of the campus, we decided to arrange a three-day international conference. We secured outstanding speakers for the general topic, "What should be the common elements of a university education in all countries of the world?"
The speakers included:

Arnold J. Toynbee, Historian
Lady Barbara Ward Jackson, Economist, Writer
Kenneth Holland, President, Institute of International Education
William Paddock, Head, Latin American Affairs, National Academy of
    Sciences
Jaime Benitez, Chancellor, University of Puerto Rico
Cleanth Brooks, Cultural Attaché, American Embassy, London
Hamden Forkner, Professor Emeritus, Teachers College, Columbia University
    sity
Sir William Hayter, Warden of New College
Arthur L. P. Norrington, Pro-Vice-Chancellor, Oxford University
Carlos Romulo, Chancellor, University of the Philippines

John Paul Getty, Financier
John Scott, Roving Correspondent, *Time* Magazine

It was the most difficult type of meeting to run because we had it partly in Oxford and partly in Wroxton, and the logistics involved were of the most complicated type. The Hotel Randolph in Oxford is very small, but we were fortunate in securing the dormitory rooms at New College since the conference was held during the summer. The sessions themselves were held at Rhodes House. We started the conference with Chapel exercises in the beautiful New College Chapel, the Lord Bishop of Oxford officiating. We marched in cap and gown from Rhodes Hall through the quaint streets of Oxford to the Chapel and then back again. The conference went without a flaw and every speaker came through beautifully and excited a great deal of discussion. We had about 250 representatives from colleges and universities from twenty-one different countries. The day before the meeting was to begin I received a message that the King of Italy wanted to speak to me. I thought it was a joke, of course, but then I remembered that some months before, at a party given by one of our trustees, I was speaking to Dick West, formerly president of the Irving Trust Company in New York. In talking about Wroxton, Mr. West had said: "Why don't you invite King Humbert. I am sure he would like to come." Both Mr. West and I had received decorations from the reigning Savoy family, the Order of the Crown of Italy, and I had commented about his decoration, which he happened to be wearing and which I had recognized. I went to the telephone; indeed it was the minister for King Humbert, who said that the King had received my invitation and would like to come to the conference but would like to have lunch with me before the conference.

Of course, I said. We planned, therefore, to meet for lunch in the dining-room of the Hotel Randolph in Oxford the following day. We arrived on time, and as I was going from the foyer to the dining-room I naturally met many of the invited people along the way. I introduced each to His Royal Highness, King Humbert, and, of course, for a moment, no one knew whether I was pulling his leg or not. It was a thoroughly enjoyable lunch except for a thoroughly execrable Manhattan cocktail, which one sometimes gets in the less worldly bars of England.

I found out later that the King had come from his place in Cascais in Portugal, where he is living in exile, to the Hotel Savoy in London, in order to attend the conference. He attended practically all of the sessions and took a very warm interest in everything that was going on. On the last day we had the official dedication with a full-fledged academic procession. The Abbey's grounds never looked lovelier and we were blessed with three days of sunshine during the course of the meetings. The procession wound its way from the Abbey through the village streets to the nearby village church. Those of you who have seen the motion picture "Tom Jones" may remember the old church with its cemetery alongside. The church looked very much like the motion picture setting. The idea was to have the official banner at Wroxton college blessed in the medieval manner by the bishop, and again the Lord Bishop of Oxford was there to head the procession, aided by Canon Edgar C. Taylor of St. Louis and the Reverend Robert E. L. Walker, Vicar of the church. Both men held up the very heavy Bishop's robes. Dr. Heinz Mackensen, chairman of the University Council, carried the university banner and I led the procession with Lord Harlech. Lila Tyng of our Board of Fellows accompanied the King of Italy. Before the procession we had had a picnic in the English style. On one of my airplane trips I had decided on every little detail for the picnic box, including a quarter bottle of Montrachet. But a curious thing happened. Some of the participants brought their children along to the picnic and before we knew it, the children had got to the boxes before the adults. The result was that those at the tail end were left without picnic boxes and many people had to share. When Lord Harlech and King Humbert arrived, there was no picnic box for them. Sally had to go scurrying around taking a sandwich from one box and stealing a banana from somebody else in order to make up a presentable meal for our honored guests.

We all crowded into the old church and the official blessing was one of the most touching ceremonies in my academic experience. We marched back to the Abbey to participate in afternoon tea, champagne, and strawberries. One of the features of the afternoon function was the appearance of the First Air Force Band, which played for the procession and also played the official march of Wroxton College that had been composed by one of our faculty members, Stanley A. Purdy.

That evening we had the gala dinner and ball. It was a white tie affair with decorations. Since the Abbey, big as it was, could not accommodate all of the guests, we had a marquee outside for the dining. Cocktails were served in the Music Room of the Abbey; our receiving line was in the Great Hall. As usual, our toastmaster was there, this time functioning in the role of a caller.

In England, banquets and other public functions are presided over not by a selected amateur but by a professional toastmaster, usually attired in red. It was this gentleman who had been the announcer at the cocktail party in the old cloisters of New College but who was now calling out the names of the guests as they began to flow into the great hall of the Abbey. He looked over the list of guests in order to familiarize himself with their names and asked me how to announce King Humbert. I replied. "His Royal Highness, King Humbert." "In England," he said, "we never give the name, since everyone knows him." As usual, I made up my own decision on protocol and requested him to stick to the full phrase and name. The King arrived with his minister, Count Olivieri. He wore a whole panoply of jeweled decorations which lent éclat to the hallowed hall. The receiving line had, in addition to Sally and me, Ed Williams, Henry and Jean Becton, and Dean and Mrs. Haberly. As each couple entered, the toastmaster announced their names in stentorian voice. There is something pleasing in having your name announced as you walk into a reception and I am sure that the experience was a happy one for those who came. When the time came for us to go in to dinner, two bagpipers led us to the tables. I had chosen the menu and the wines myself. The first course was prosciutto with melon. Since the caterer had prepared dinner at various royal households, I assumed he was sophisticated in all matters of international cuisine. I looked forward to a full slice of ripe melon accompanied by a few slices of prosciutto, which is like a Westphalian ham. Imagine my surprise when what was served was a small one-inch square of melon entirely wrapped with ham and topped with my pet abomination, a maraschino cherry. Oh well!

The rest of the meal was, fortunately, more professionally standard, and it ended with a pastry chef d'oeuvre representing an American flag. I liked the assemblage. It was a convivial and happy one, and what especially pleased me was that many of our faculty were there too. Even those students who were attending our summer session were invited.

Knowing how chilly English nights can be, I had taken the precaution of having three bonfires arranged in the rear gardens. At the conclusion of the dinner, the guests found the bonfires an attractive experience. As George Gobel might say, "You hardly don't have bonfires no more." While they were taking the air, the tables were being taken away to provide for dancing. It was a grand evening in a grand manner, and when we all intertwined our hands and sang "Auld Lang Syne" we all felt that Wroxton Abbey had really been broken in as an integral part of Fairleigh Dickinson University.

19

Student Experiences

"But Dr. Sammartino, I tell you we heard the ghost moaning."

This was a spokesman for a group of Rutherford students who had stayed over on a Saturday night to hear Mrs. Ivison's ghost in the old castle.

"Nonsense," I replied. "That was your imagination."

Actually, it wasn't imagination. They actually did hear a voice. It was the first year of the college. All we had was the old castle. But there were recurring stories around Rutherford of Mrs. Ivison's ghost that came back once a year on the anniversary of her death to revisit her home at midnight to seek her husband, David. From the records of nearby Grace Episcopal Church, the students had ascertained the date of her demise, which happened to fall on a Saturday night. They had come in to ask permission to stay over until midnight and scotch the silly story. In those days, this was the extent of student sit-ins. Reluctantly, I gave them permission and then began to realize that granting such a simple request did have complications. For one thing, this was a war year and the heat was reduced during weekends. Second, we had no custodian on duty. Third, in those days administrators were very careful to preserve the chastity of their students and to leave a group of boys and girls alone in the castle did not jibe with the Victorian standards that guided our administrative procedures. The young man in charge of evening activities, Julius Luck, happened to pass by. I discussed the situation with him. As I remember it, we thought all the students might just as well have a party ending at about eleven and those wishing to wait for the midnight arrival of the ghost could stay over. That would leave one hour or so for the troublesome period. I asked Julie whether he would stay over and lock up the castle, and then a thought came to me. Why not have some fun and provide the semblance of a ghost? Julie was a young bachelor in those days and he fell in gleefully with the idea. As we discussed the idea, we embellished it like two little boys cooking up a story. On the third floor there was a secret passage from one room to another. Julie could chaperone the party and then he could make believe he was going home but actually come in unseen through the rear porte-cochère, then go up quietly to the third floor and hide in the secret passage. Above the second-floor grand staircase there was a frosted-glass light well. At the stroke of midnight, with all lights out, Julie was to tiptoe out, play a flashlight on the frosted glass, and moan "David! David!" He was to go back quietly to the secret passageway, so that if the students started to look, they would find no one.

As I got the story later, the students ended their party and most of them went home, but a dozen or so stayed and sat quietly in the eerie gloom of the castle. At midnight, the faint lights began to flicker through the frosted glass. The barely audible words "David, David," wafted over the empty stairwell. One girl screamed; most of the students rushed out. Two of the boys, braver than the rest, rushed upstairs to see if anyone was there. No one! Was there really a ghost? The students really became convinced there was. One, Winston Rockefeller, said to me on the Monday morning,

"I can't figure it out. We're all sure we saw a faint light form. We're all sure we heard the words, even though it was a whisper. And yet, we looked in every room above the first floor and no one was there."

"Oh, you were just imagining things," I said hypocritically.

"Oh, no; we're all nonbelievers in ghosts and yet, and yet. . . ." Winston said.

The legend was to perpetuate itself in the name of a student magazine. But as other buildings were erected around the castle, little by little the romantic aura of the castle lessened. Students used to clear out the great classroom and have their dances in the original salon of the castle. They prepared their own refreshments in the nearby kitchen and worked out their own decorative schemes. The young women used to love to come down the grand staircase in queenly fashion and be received by the men, who waited patiently at its foot. But as the war really got started, men were scarce. What to do? We conceived the idea of inviting seniors from the New York Military Academy. After all, the age difference was not too much. Through Colonel Dickinson I had come to know Colonel Pattillo, the head of the Academy, and he was very cooperative. A busload of apple-cheeked cadets arrived one Saturday night in their spanking new uniforms, ready to fill in for the men who were away at war. Alas, the evening was a social failure. There is no difference quite so great as a year at age eighteen. The girls looked down on the boys as too young and immature. As they admitted later, they felt they were coddling their young kid brothers. Actually, I couldn't see that much difference. In the absence of men in general, we assumed the girls would be glad to have male companionship. When I talked with Colonel Pattillo later, he told me his men had had a wonderful time. They felt they had really wowed the girls and were anxious to come back.

The next time, we thought we might ask the officers and enlisted men from Camp Shanks, which was nearby. The commanding officer was only too glad to oblige and thought it was a good idea for some of his men to attend a social function "comme il faut" at a nearby college. He provided two buses for them. Came Saturday night and the girls were all dressed up in their finest, waiting in the castle. Came the time for the arrival of the men—and no men. A frantic telephone call to the camp revealed that they had left hours ago. Finally the two busses pulled up in the parking lot. The men got out, with their hats turned around to simulate German sailors. The formed ranks and then started to goose-step toward the castle, chanting "Ein, zwei, drei, vier, Achtung!" Let me explain that while Rutherford itself is a dry town, East Rutherford across the tracks is anything but dry. The men had stopped in East Rutherford for refreshments and by the time they reached the college campus, they were a well-oiled group of soldiers. The girls looked out at the inebriated army with terrified faces. Sally and I have always been able to act together on a split second basis.

"Get some postcards of the castle," I told her. While she was getting them, I said to the girls,

"This is a test to see how you can handle men. Start telling them about the college. Have them write cards to their families. That way you can find out who is married and who is single (titters). Keep their minds on their families and mother them along."

I was proud of the girls that evening. They handled the men like mature women. They danced the men around until they were ready to drop. When one man threatened to become too attentive, another girl or two would break in and lead his attention elsewhere. In some unaccountable way, we got through the evening without any sad results. The girls enjoyed the challenge. The men seemed to have had a good time. At eleven, they all stood in one great circle, crossed and clasped their hands, and sang "Auld Lang Syne." The busses pulled up and the men and girls waved goodbye to each other. What had threatened to become a raucous bacchanale turned out to be a perfectly wholesome evening. By this time some of the parents had come up for their daughters. It was wartime and cars were a scarcity. I called the girls together and said:

"Girls, I am proud of you."

I explained to the parents what had happened and went on, "You've had one of the greatest experiences of your life. You took a group of seemingly unruly men bent on no good and you handled them beautifully. You didn't flinch. You were complete masters of the situation and, what's more, you had a good time to boot." I noticed two mothers with tears streaming down their cheeks. It was an evening, to say the least. As I was going out, the custodian came to me with four hats that had been left behind.

"What'll I do with these?"

"Just keep them in a closet. If no one calls for them, some of the girls might like to wear them in memory of a glorious evening."

Over the years I have had students drive for me. They didn't make much money but it was the kind of job that left a great deal of time for reading while I was attending a meeting or making a speech. Many times they would meet important guests at the airports and they were trained to fill in the guests on Fairleigh Dickinson University along the way and answer any questions. They got to know many important people and thoroughly enjoyed the experience. But I learned a great deal from these young men. During the late fifties and early sixties I became concerned about the mores of young people. The reports from other colleges were as distressing as ours. One young man suggested that a group of men and women students meet with me so that I could learn directly from them. He was a student leader and easily formed a committee of eight who used to meet with me regularly so that I, the president of the college, could know what was happening. I learned a great deal from them, much of which I was not prepared to accept or to understand. But, in short, what they said was that their lives were their own affair. The day of parental or institutional control over their sex life was over. The pill had not come into being at that time, but the proper use of contraceptives made sex activities relatively safe. They admitted that at times they became careless and in most cases any resultant pregnancies led to hurried marriages. It meant a hastening of plans, but one way to married bliss was as good as another. There was a strong feeling on the part of the coeds to have the same freedom the man has without the stigma of social disapproval. In spite of the then feeling that Victorian adminstrative watchfulness was a thing of the past, they did not resent our stringent house rules. They would have their gay times regardless of rules and the whole matter was treated as a sort of game. At times they would laugh at my naïveté, but there was always a feeling of camaraderie. I was taken aback by some of their ideas but I never felt that they resented my stiffer attitudes. Often we would have closed-ballot votes in order not to embarrass anyone, but it was not really necessary. The girls were just as outspoken as the men. They deplored promiscuity, but it seemed to me that the men were ready to pluck whenever the opportunity arose. What is the answer to all this? There is no easy solution. One of the factors we overlook is that many parents don't object to a little promiscuity if it's a prelude to a good marriage. Most parents don't object to the sowing af a few wild oats by their sons. With daughters, they are a little more squeamish. But I have had some frank and open discussions with mothers at cocktail parties and, curiously, I find that mothers are more permissive than fathers regarding their daughter's sexual experiences. Is this because fathers understand more the hunter spirit of men, or is it because mothers understand more the needs of women? But many mothers do not resent discreet experiences by their daughters if they have a reasonable chance of leading to marriage. But it must be discreet and, of course, with proper precautions to prevent pregnancy.

In spite of all of the above, while I can't prove it, I think the majority of young people would just as soon have regulated dormitories. It's no fun, and indeed it can have serious psychological effects on certain individuals, to live in a free-for-all dormitory where anything goes and la dolce vita becomes the prevailing standard. Will this point of view win out? I don't think so. Administrators are tired of being cops, and cops can function effectively only as long as they have a reasonable chance of holding the line.

The day of the institutional dormitory may be over. Students take the point of view that, since they are paying for the room they occupy, they can do anything they want to. In one institution one student was letting other members of his family occupy his quarters and this probably takes place in other institutions. The rented room becomes just that and nothing more. It really doesn't make any sense any more for institutions to be in the rooming business. It used to have a purpose when the college was helping young people to preserve their virginity and to help them learn to live like human beings. The college was in loco parentis. Now it no longer is, and there is no more reason for a college to worry about rooms than for it to run a haberdashery for the men or a dress shop for the women.

20

The Athletic Rat Race

My *secretary ushered in* one of our basketball players, on crutches, and his mother, whose eyes were red from crying. Laboriously, the young man arranged his long legs, one of which was in a cast, so that he could sit at the tip of his chair. The young man was a fairly good student and an outstanding ball-player to boot. He was a pleasant chap, not particularly outgoing, but well liked by the faculty and students. I couldn't imagine what the problem was. He started to speak haltingly:

"I've broken my leg."

"I can see that," I said, laughing.

"Yes, but now I can't play basketball."

"You'd better not." I said, still in a light vein.

"Yes, but now I'll lose my basketball scholarship and, since I don't have money for my tuition, I'll have to drop out of college." Then his mother started to cry and tears welled up in the young man's eyes.

"Who said you were going to lose your scholarship? You got your scholarship because of two reasons: one, you were in financial need and could not have gone to college without help, and second, your grades in high school were very good. The fact that you played basketball was of secondary importance. You can't play this season; we'll miss you on the team but you don't lose your scholarship: What's the problem?"

"I can stay in school?"

"Of course you can!"

At this point his mother came over and, bawling louder than ever, came over to kiss my hands. The young man was now crying like a baby and smiling at the same time. It was one of the most ludicrous scenes I had ever participated in, but in a few moments all was sweetness and happiness. I explained to the young man that our rules were different from those in most sport-oriented colleges. Once we encouraged a student to embark on an academic career with a scholarship, we felt we had a moral obligation to see him through to graduation unless he, through lack of concentration on his academic work, fell down on his grades. It was as simple as that.

When Fairleigh Dickinson was founded in 1942, I was as uninterested in intercollegiate athletics as anyone could be. What had started in the nineteenth century as a purely amateur and friendly contest had now become an expensive, hypocritical, and almost professional rat race which to my mind had very little to offer an academic institution. I felt strongly that emphasis should be placed on lifetime sports: swimming, golf, tennis, bowling, skiing, and even dancing. Students in college, boys more so than girls, are still in a period when team sports have some value and, to appease this spirit, there were teams in all sports except football. But to give you an idea of how the lifetime sport worked, at a meeting of all students (we were small enough to be able to do this), I had the pro from my own golf club give a demonstration on stage of the proper stance and the proper way to hit a ball, using plastic balls which he would hit right into the audience. Then a few students, including girls, would be coaxed to come up and try, the pro pointing out any errors. Then there would be shown a map indicating all the public courses in the area. Finally, we indicated that to start with, it was unnecessary to buy a complete set of golf clubs. We had a minimum-priced set featured in the bookstore, which included only two woods, a three, five, seven, and nine irons, and a putter. Students could then sign up for a short eight-session course in golf, ten in a group. We established as many groups as requested, boys and girls together. Those who wanted an advanced course could have it. There was no rigmarole about credit, about a semester-long course, about tests. We were trying to get young men and women to enjoy an exhilarating sport together, to make maximum use of community facilities at minimum cost, and to give them something they could enjoy all their lives.

Now to come to team sports. If they are kept on a nonserious, casual level they have some value: young people are brought together, there is a certain amount of horse play, the activity does take on a social nature. No great heroes are born and lesser players are encouraged to try their best. But is this what regular intercollegiate athletics comes to? No! I used to go to the basketball games. I was appalled at the spirit at these games. The idea was to win at any cost, even if you had to cheat. The referee was a crook and a dunce if he made a decision against you. And what is even more perplexing is that if the team has a winning streak college spirit is high. Let it lose three or four games in a row, attendance drops perceptibly and the spirit flags—this at a time when the team needs cheering most. I suggested at one time that our own team cheer any good plays by the opposition. Then they in turn would have to cheer good plays by us. (The Chinese actually did just that during the 1971 ping-pong tournament in Red China, applauding each time a member of the United States team made a point.) The emphasis would then be on good playing. The students shrieked in horror. This would be pure treason. I thought baseball surely would have a sportsmanlike aura about it. It was even worse. The idea was to rattle the opposing players by any means, usually foul. I thought girls' fencing would surely be a pleasantly played activity. This was the worst of all. I am sure that if it had been possible to gouge out the eyes of the opposing players and get away with it, it would have been done. As far as I can see, about the only two intercollegiate sports where there is no hanky-panky and where a tolerable spirit of fair play exists are track and golf. I don't believe we are doing young people any good when we encourage cheating, foul play, discourtesy, and outright mass belligerency.

Now, of course, if the college team becomes so good that it gets to the top nationally, then you wind up with a near-professional activity. It is no longer an amateur sport and indirectly the institution is buying its players. At Fairleigh Dickinson, as long as I had my hand on the pulse of things, I always insisted that athletic scholarships be given on the same basis as any other scholarships: academic standing and financial need.

Getting back to basketball, it is not a sport for normal beings. It is practically restricted to very tall players; it has become a game for the abnormally tall. Almost as a gag I was going to bring to the college five Watusi from Africa. Giraffe-like, they could keep the ball on a higher plane and practically drop the ball through the hoop at will. It was too difficult to find enough with the proper academic qualifications. I proposed at one time to have two classes of basketball; one for the very tall and one for the normal-height students. We do much the same in boxing. This idea came to naught also.

A few, but very few out of the 2500 colleges in America, do make enough from football to pay for the other sports. What is really happening is that the institution is in a sort of business to pay for its other sports.

During the early days of the college, I used to consult often with the high school principals in northern New Jersey. They would say with great determination, "Peter, don't have a football team. It will dilute your academic effort. It will squander money on a sport a comparatively small number of students will play. You will have to build a winning team if you expect students and the public to come to the games to be entertained. You will soon become enmeshed in the rat race of competing for athletes and the concomitant bribes that will be necessary to attract them." Football is a rat race no matter how you look at it. It does build up for a few fleeting hours an artificial and sophomoric school spirit, and I suppose in the nature of things this does have a certain value. I should think that the emphasis ought to be on building up an élan based on things of the mind or on serving other people. There was a time back in 1893 when the amicable contest between Yale and Princeton had New York in a tizzy. The Vanderbilt and Whitney mansions flew the Yale banner. Sloanes and Scribners vaunted the Princeton flag. Games became more brutal each year and, in 1905, eighteen players died because of football. Believe it or not, President Roosevelt called the coaches together at the White House and told them, "Get the game played on a thoroughly clean basis." Some colleges gave up football, including Columbia, California, and Stanford. But when Eliot tried to abolish it at Harvard, Roosevelt said in a speech, "I think Harvard will be doing a baby act if she takes any such foolish course." Rah-rah-rah! Today, pro-football seems to be stealing the rah-rah from colleges and making it available to everybody—at a price, of course.

Through the years, pressure groups of students have tried to establish football at Fairleigh Dickinson. I always brought up their request at a convocation of students and explained the economics of football. It would cost each student $50 extra a year to support football. How many were ready to pay it? Very few hands were raised. Now we have come full cycle with Club Football. Everything is now supposed to be back on the nice, easygoing, inexpensive nonprofessional level. The students themselves are supposed to be running the show. No bigtime games, no highly paid coaches, and no scholarships. What the result will be, I don't know. But I do know that some students have expressed the thought that if Club Football is introduced, the next step would be regular football.

The role of physical training during college is another expense that most students have to bear whether they like it or not. To get hordes of students two or three times a week to do exercises doesn't seem to make much sense to me. This doesn't mean I don't believe in exercise. On the contrary, when I was a classroom teacher, before I started my classes in French, I would open the windows wide, have the students do some deep breathing, and then do a few minutes of exercise. The truth was that I wanted the exercise for myself but if it was good for me it was good for the students too. It was invigorating and it sharpened our minds. I would be in favor of simple exercises done two or three times a day for a five-minute stretch each time so as to get the blood circulating and our bodies lithe. But by the time students get to college, the expense of the physical training program is not justified by the results obtained. But I would have special classes for those who wish them five days a week if necessary.

There is another factor to be considered. If a student is in a dormitory college and has no outside job, the problem is different from that of commuting students or even of a boarding student who may be waiting on tables or already participating in sports. There is no point to the commuting student's driving up madly to the parking lot, rushing to change into gym clothes for twenty minutes or so of exercise, and then rushing back to a job. I might add that we have not given isometric exercises the prominence they deserve, nor have we stressed the daily walking that could be done. The student drives to the college gymnasium where he is supposed to overcome the stiffness caused by driving.

The role of physical training and athletics should be a harmonious one, serving the physical, social, and emotional needs of the individual. It has to integrate with the overall program of the individual, including work and study schedules, and it should be involved with the total health of the person. Not least in importance are the social aspects of such activities. A college is not in the business of training highly polished athletes. It is not a school for gladiators as in the days of ancient Rome. Nor is it a professional training school for professional players of the present-day period.

People little realize the financial cost of keeping up with the Joneses in the academic community. We are spending millions to accomplish relatively little while we eschew more natural and much less expensive methods of physical development. To give an example: I know of a college for two thousand students that has a gymnasium costing three million dollars. Now, the amortization and operational expense of such a gymnasium is conservatively about twenty percent, or $600,000 a year. Divided by 2,000 students, this means an expense of $300 per student per year. And this does not include faculty expense. Now suppose we said to a student: look, your tuition is $1000 a year but for $300 extra you can have a gymnasium to play around in. For $100 extra you can have the use of an outside field. For another $50 you can use our tennis courts and for $200 our golf course. On that basis, how many of our students—or parents, who are in most cases paying the bills—would be ready to pay for the extra services? To go a step further, how many students would be willing, as we have already shown, to pay extra to subsidize intercollegiate sports for a relatively small group of athletes? A friend of mine used to play polo when he went to Yale. But each player bought and paid for his own horse. He didn't expect Yale to subsidize his sport. Rationally, it would make better sense to charge each player for his own participation in any sport, just as we do in life. If we can afford to, we belong to a golf club. If we go bowling, we pay the going rate. But one reason school costs are skyrocketing is because of the tremendous expense of supporting intercollegiate teams and for the equally large expenditures in supporting a physical training establishment whose returns in student health are not easily discernible.

As I am writing this, New York University has decided to drop basketball and track because of the costs involved. It makes sense. In these days of greatly increased tuition costs, why add to the students' tuition a subsidy for teams that serve a relatively small number of students? The lesson is there for all to see. If you want a splash in intercollegiate athletics, you have to subsidize teams, and this costs money. Now, if a large institution such as New York University finds the process painful, imagine what it must be for a small private college. After all, such a college has to have just as many people on the basketball squad as a large one, but the cost is shared by a much smaller number of students.

In public institutions the cost is usually borne by taxpayers and taxpayers are grumbling about high taxes. Fortunately, this problem can be solved easily. Cut out the mumbo-jumbo about intercollegiate athletics and concentrate on low-cost intramural and lifetime sports.

On January 29, 1958, I received a letter from an unknown person with the odd-sounding name, say, of Lavinia Pooch. I can assure you her real name was just as odd. The letter started off with these paragraphs:

"Because I am unable to come to see you, I am writing this letter to ask you if I provide the expenses of the trip here, will you and Mrs. Sammartino come to see me?

"My reasons for wanting to interview someone from Fairleigh Dickinson University are many and somewhat complicated. I have a copy of your catalogue, and together with other pertinent information, have chosen this school as the beneficiary of my endeavors.

"What week-end could you get away? By train from Newark, I've found, is the quickest way to Washington. Or perhaps you prefer to drive here. Whatever expenses would incur, I insist upon assuming. If one or both of you feel you cannot get away, please appoint someone in your place to come. However, it is you and Mrs. Sammartino that *should* come.

"Please do not write and ask me . . . to be more explicit about the *"why"* . . . of this request. My answer would be as complicated as yours would have been, had someone asked you. . . . "Why did you want to be a pioneer? . . . "

I showed the letter to Ed Williams, Chairman of the Board of Trustees, and he agreed that Sally and I should make the trip to Washington and not lose this great opportunity.

I wrote to Mrs. Pooch and told her which train we would be arriving on. I expected to be met with an old-fashioned black limousine, brightly polished and with a uniformed chauffeur, perhaps even an extra footman. Well, the car that met us was old-fashioned all right, an old-fashioned station wagon, ready to fall apart and with cracked window panes held together with paper tape. Sally and I both gulped, but we hoped that this was simply an idiosyncrasy of an old lady—of a rich old lady. We drove to the "estate" expecting a sprawling domain with circular drives, porte cochère, and servants running out to meet us. Indeed, along the way, there were continuous references to the large estates we passed, although the conversation centered on horses and fox hunts. Instead, we arrived at a stone house, amateurishly put together in a "do-it-yourself" fashion. In her letter Mrs. Pooch had mentioned a fireplace built of thirty tons of field rock. The stone fireplace was as large as she had said it was, but was outsize in an ugly fashion and out of all proportion to the demure living room. At this point we needed a drink badly. This was, however, one of those homes where liquor is never served.

We sat down to lunch trying to figure out what it was all about. We had what our hostess kept referring to as "s'ghetti," or spaghetti with tomato sauce. Probably meat would have been too expensive. Everything was very much "en famille" and Sally wound up helping with the dishes in a small and decidedly old-fashioned and disorderly kitchen.

Finally, with everything in order, we sat around in the diminutive living room with the fifteen-foot fireplace gaping at us, and awaited tensely for the great moment to arrive. What exactly was the great opportunity for us?

Well, it seemed that Mrs. Pooch had written a film script on how students should solve their problems. She had tried to place it with one station after another but none of them was smart enough to appreciate the gem she was offering them. She was sure that once the film was produced, the royalties would just roll in. Her proposal was simple. If our college produced the film, it would show the world how perceptive we were in seizing upon this gem and we would participate to the extent of hundreds of thousands of dollars.

This time I really gulped. I almost burst out laughing. The situation was really ludicrous. I didn't laugh because I felt sorry for the lady. Senseless as it seemed to me, as far as she was concerned she was a genius with a film that was to lead humanity. As I remember, I did some lightning calculations in my head. I knew that there was a train from Washington at about four o'clock and answered as follows:

"First of all, I have to go back for an alumni dinner tonight and I don't want to miss the four o'clock train. Can I get a taxi to take me back in time?" I was assured that this would be no problem.

I continued. "I'm sure your film is a very worthwhile one. I do not have the power to decide on what is produced. I'll be glad to have your manuscript read by the faculty committee and we shall inform you in due time." How much more pious can you get under such circumstances? The air was somewhat artificial. We all made up phrases that didn't have too much substance. Mr. Pooch hustled off to call a taxi, for which I was very grateful.

"May I take the manuscript with me?" I said in an impulsive way, in order not to let the lady down too much. The wait for the taxi was one of the longest I have experienced. Mercifully it came at last. We both scooted out, thanking our hostess for the wonderful "S'ghetti."

Once in the taxi and safely out of sight, Sally and I broke out into broad smiles that said: "Boy! Have we been took! Oh well, c'est la vie. It was too good to be true."

On the train, luckily, there was a dining car. While we had our much-delayed drinks, we glanced at the priceless manuscript and felt sorry for the lady who was to shake the world.

Like most college presidents, I found that fund raising was the part of the job I liked least, and yet some of the greatest friendships in our lives have stemmed from contacts that at the beginning were primarily for fund raising. By the time I retired, practically all the donors to the university had become close and cherished friends and I did not feel that their friendship depended on whether they contributed or not. And, because Sally and I contributed our mite to the institution, we felt less and less like mendicants asking for alms for the university. Indeed, with the exception of five of the top contributors, whose fortunes were ten or a hundred times that of ours, we were not asking prospects to do anything we weren't already doing. This attitude can have a tremendous difference in your asking. I always felt that I was doing the other person a favor by giving him an opportunity to donate to a worthy cause. Sometimes the situations were almost humorous. In one case, the gentleman in question, who had retired with over a million dollars, said magnanimously that he would give me $200, and then added as a suggestion, since time hung heavy on his hands, that I might visit him every week. At the current rate of baby-sitters, I would just about earn the money he was contributing. I never found time to visit him and I never got any more contributions.

Curiously, after the first request for help from Colonel Fairleigh S. Dickinson, I never asked him for any money. He always gave it of his own accord and the same holds true of his son Dick, with one exception—his gift to the Dental School, which I have already recounted.

Not all money-raising visits were so felicitous. Once I called on a Mr. Brown, who had been a successful industrialist and had built up his own company from nothing. When I suggested a modest gift to the university, he started on a tirade against colleges in general, what a disservice they were doing humanity, the awful conditions on campuses, and how he had built his company without the benefit of a college education. It didn't sound as if I was going to get any money on this trip. Finally he turned to me and said,

"Well, what do you think of me, the president of my company?"

I couldn't resist a bon mot which had been hanging on my lips for some months:

"Mr. Brown, I think you are non campus mentis."

Well, that clinched my zero average for the day. Nevertheless I went out with a self-satisfied smile on my face. Some time later, I met Mr. Brown at the golf club and he asked me what I meant by the remark. I explained the pun to him, which he now understood, and we laughed together. A few weeks later he sent me a check for $500.

I never really worried too much about raising money. I felt that it made much better sense to give my main energy to running the college. With day and evening sessions and summer classes, I had enough to keep me busy day and night twelve months of the year. Let me give an example of what I mean. In one instance, in the forties, when our college was small, I had to make a decision as to whether to have one or two custodians in a building. The dean and I were the only two administrators. It was up to me to decide. There were no norms to go by and whatever information I could get seemed irrelevant. So, one night, when the last evening class was out, I slipped into overalls, took a push broom, and timed myself on cleaning three rooms. As I remember it, I did the work in less than an hour. It was apparent that for this building one custodian was more than enough. The experience was to stand me in good stead for many years after that.

I didn't want to waste endless hours getting a hundred dollars here, two hundred there. On the other hand, if a person seemed interested in the college, had the means of making a major contribution, and, most important, had an enlightened community point of view, then I felt my expenditure of time made sense. But I solved the problem of keeping contacts with donors, large or small, in a way that blended with our living pattern, and which turned out to be pleasurable for us and, I feel sure, for the donors too. One rule I had was that wives were always included in all invitations, and this for two reasons: first, a contribution comes from the family; second, there is no good reason for taking a man from the companionship of his wife to attend a meeting that does not involve state secrets. The result was that the women felt just as much part of the college family as the men. As a group they enjoyed each other's company and the meetings and get-togethers were always pleasant. The curious thing was that these friendships had an international flavor, and whether it was at the dedication of our Wroxton campus in England or at the triennial meeting in Seoul, Korea, of the International Association of University Presidents, of which I happened to be president, there was always a good sprinkling of members of the Board of Fellows of the University. I kept up my activity because I liked people and did not think in terms of fund raising. And pleasant experiences arose from time to time. Once I was visiting a Fellow in the hospital. I was genuinely happy that his operation had been a success. As we conversed gaily, he took out a pen and gave me a check for $5,000. In another instance, also during a purely social visit, I happened to mention the problem of getting more facilities for students, and again to my surprise, for I had no ulterior motive, this friend then and there wrote out a check for $25,000. In still another instance—but in this case the gentleman had asked me whether he could help a student or two—I gave the name of two students and the friend immediately arranged to give them partial scholarships, adding, "I'm an old man; my wife is somewhat younger. When I pass away, the income of my estate will go to her support. But at her death, the estate will go to the college and it will probably be worth about $300,000."

Another example of how gifts may happen purely by chance was one that led to the founding of an experimental college. From time to time I asked for a special informal meeting of the Board of Trustees. At this particular one, I explained that I felt that in ten or fifteen years higher education would break into four segments:

the first two years; the next three years leading directly to the Master's degree; then the three years or so leading to the doctorate; and finally, the post-doctoral area. More and more students were making the master's instead of the bachelor's degree their terminal degree. I told them we should experiment with a small college limited to 400 students, and if that were successful, put all of our first two years into groupings of 400 or so. The meeting was a stimulating one but was purely on a "think" level. Two weeks later Mr. Dickinson came over to see me and said, "Peter, I like your idea and if you want to put up the building I'll give you 20,000 shares of Becton-Dickinson stock worth approximately $1,000,000." It was that simple, but we still had to present the idea to the University Council and while technically we had the right to do anything we wanted, actually we would not have established the new college if the faculty would not support the idea. We asked for a special meeting of the Council and, as it happened, the project was accepted unanimously. A faculty committee was appointed by the chairman to work out the curriculum for the experimental two-year college. The curriculum was developed and was then presented to the Educational Policies Committee and duly approved. In the meantime, I was in almost daily touch with our architect, Roland Wank, since I wanted this building to throw aside the traditional institutional aspects and be designed to meet a new concept. Without increasing costs more than five percent, I wanted to create an atmosphere of a mature pleasant center for students—a place of joy and learning where human friendships among students and their faculty would have an opportunity to flourish, an ambiance permeated with good conversation and a zest for intellectual growth. I insisted on wall-to-wall carpeting in the classrooms and in the halls. I wanted the furniture different in each room and of the type one would see in an outstanding lounge room in contemporary styling. I then set about having a sort of little art gallery in each room. In some cases I got friends to contribute paintings; in other cases I bought posters or maps. For two rooms, I brought in paintings at my own expense. I also wanted to do away with that awful barrier between faculty and students—the desk. The desk was never intended for the classroom. It just slipped in there when no one was looking. The desk is meant for the office, or for the library or study. In the classroom it has acquired two functions, both of them bad. It separates the teachers from the students, and it becomes a catchall for all the junk a faculty member can't decide whether to throw away or not. Besides being behind a desk, some faculty members have a tendency to look down at their notes, which decreases their hold on their audience all the more. Some male instructors try to reach their listeners by sitting on the edge of their desks. This is hardly a dignified professional position, but many carry it off with aplomb. Female instructors rarely adopt this position and, of course, if they did, with the advent of the miniskirt it would have the effect of attracting the students' gaze to areas other than that from which the golden words pour forth.

I tried without success to substitute a coffee table for a desk. This was low enough so that the conversational flow would be easier and natural. A coffee table is hardly a barrier and yet the instructor can lay his notes on them and refer to them from time to time. No go. An instructor just isn't used to such new classroom habits. Besides, here and there an instructor even used it as a footstool. Finally we compromised on simple tables—and no drawers. We did put a two-drawer file in each room. By actual observation and experimentation, I came to the conclusion that no instructor really needs more than one file-drawer for his classroom. This can hold all the papers he needs at hand, plus pipes, lunch, and thermos bottle. In an unusual case where for some very special reason he needs more room, the exception can always be made.

I designed a special high chair so that the instructor could be above the students and encourage attention. I got this idea from author Kenneth Roberts, who, in his cocktail room, had a high chair which he always occupied. When he spoke he liked to have people pay attention to what he was saying. My chair had a tablet arm for the instructor's notes, and it could swivel. I felt it was a good device to prevent the faculty member from poring over his notes and give him greater mobility in the classroom.

Although the students could use the large library across the river in Teaneck, I had a special small library for them, and it was a jewel. Any book—within reason, of course—that the students or faculty wanted, we put in it. We wanted it to be a joyful place that would encourage browsing and reading for knowledge and pleasure.

We didn't have the room to do what is done at the University Club in New York, have many large tables with some books on tilted stands, others lying flat, so that as a person glides past the table, his eyes are attracted by one enticing volume after another. I have never gone past such tables without picking up some volumes and winding up with at least one book that I decided to read. This to me is the ideal way of presenting books to people. Alas, it takes too much space, is difficult to administer, is inefficient to operate, and it probably encourages "permanent borrowing." Publishers spend millions to have attractive book covers, practically all of which are ripped off in libraries. But at least at Edward Williams, I won my battle to have most of the book jackets retained.

I also wanted a small auditorium that would be the community learning place for the whole college. There are many hours during the week when all the students can gather to listen to lectures, to argue with outside speakers, to have their own self-propelled discussions. It was a delightful and beautiful place with warmth and color. And on the side wall, since there would be many discussions of international matters, I had a good artist paint a mural that was really a huge map of the world. I remember when my good friend Ambassador Alimadi of Uganda came to speak to the students about the trials of a new country, most of the students really had no exact idea of where the country was located. I gave a pointer to the Ambassador and waved my hand toward the side wall. In the most natural manner, as if he had been doing it for years, Mr. Alimadi went over to the mural and pointed to the place in Africa where his country was located. This was to be repeated at future meetings, all of which goes to show that a small auditorium, cozily designed, with educational aids unobtrusively at hand, can become a most organic setting for learning.

We decided also to do other things differently. The tuition, which incidentally was less than in the University, would include all textbooks and required reading and tickets to all out-of-class activities—theatre, symphony, opera, and ballet. Too often I have seen students skimping on textbooks or some other book they were supposed to buy. The books one reads during college days should become, in a way, the cornerstone of a life library. The student should be able to mark them, to make marginal notes. This he cannot or should not do to library books. Besides, with the advent of the paperback, the whole concept of required reading is changed. A book is a precious thing. Possessing it makes it part of one's life and this is what we wanted to encourage the students to do. Suddenly the whole mechanics of textbook and required reading became very simple and no bookstore was needed.

Regarding the cultural functions, again we didn't want to leave such participation to chance. If the entire student body was to see a certain play, the tickets were bought en masse. If we could get to one of the previews, so much the better. Then the students could write their own critique and compare it with the reviews that came out in the *New York Times.* Sometimes they disagreed with the critics and in one case, as in *The Owl and the Pussycat,* they invited him to battle it out in the auditorium, which he did. It all made for serious and pleasurable theatergoing. When Sally and I attended the Vivian Beaumont Theatre in Lincoln Center, we would usually come upon a group of Edward Williams students (it wasn't always possible to get them all in on one performance) who were excitedly arguing the merits of the play with their instructors. And it was good for the instructors too, because too many of them tend to leave for tomorrow-and-tomorrow participation in the cultural events that should be part of the life of the academic man. It was the same for the opera, the symphony, and the ballet. For many of them it was the first experience in all of these activities. It gave the students an opportunity to converse intelligently and in a sophisticated manner: it was a socializing experience too. What can be more pleasurable to a girl than to go to a symphony and sit between two male classmates, and then discuss the performance? Even if the students don't like it, they have something in common and it gives them a yardstick for future happenings.

Good writing and lots of reading were expected in all classes, not just once in a while, in the English department. Effective speaking as a part of everyday experience was encouraged. You can do these things when your college group is small, when every student knows every other student, where every faculty member is intensely interested in all his students, where the emphasis is on effective living. When out-of-class visits had to be arranged, whether cultural or sociological, there were no administrative problems. The buses simply drew up to the college and there was no interference with the class schedule. This was the educational activity of the day. But above all we tried, and, I believe, succeeded in, encouraging friendliness and consideration on the part of the students. Human contacts were easy and natural and there were no lonely people. The college was an organic whole.

I think my greatest reward for the Edward Williams College idea came from a short little girl, somewhat unattractive, the kind of student that would be lost in any college class. Her father was a barber and the family lived in very Spartan surroundings. She wanted to see me on my next visit to the College.

"Doctor Sammartino," she said, "I can't tell you what this college has meant to me. If anybody had told me a year ago I would be having these experiences, I would have said 'fairytale.' I have entered a new world. I feel I have grown up and my parents are so happy."

The International Scene

We were driving back from a summer party in Sorrento, Maine, and as we turned into the driveway leading to our house in East Sullivan we saw a light in the house of a neighbor whom we knew to be away. We stopped to investigate; inside the house there was a black couple, the wife sobbing convulsively and the husband looking on helplessly. And then we suddenly realized that this was the Ambassador and his wife whom we had invited to use the lodge next to the beautiful Frenchman's Bay. They had misread the directions and had stopped at the wrong house, found no one there, and had broken a window pane to get in. But the house had been unused for almost a year and, as summer houses are apt to get, it was musty and dirty, and cluttered with outdoor furniture. The Ambassador was relatively new to this country. He had just got married and, rather than drive up in his chauffered limousine and in order to be alone on their honeymoon, they had come up by bus. Somehow he mistook my directions and now they were in a house that was distinctly unready for a new bride. No wonder she was crying. Wasn't it lucky we came along at the right time? Imagine what an international incident it could have become if a state trooper had picked up a foreign couple breaking into a private house!

We quickly took them to the lodge, which was spick and span, with beds made and even a starter breakfast in the refrigerator. No hotel could have had a nicer set-up. I have already mentioned that my friend Milton Ignatius had given the lodge to the University to be used for United Nations' Ambassadors. The bride cheered up, we had a highball to celebrate their marriage, and then we left them to their privacy.

The next evening we learned the Ambassador was indisposed. He was sleepy and felt languid. At the concert in nearby Hancock we happened to see Dr. Suyama. Could he come over and help out in an international situation? "Why, of course," he replied. And so, late that night, we dropped in at the lodge to see how the drowsy ambassador felt. Dr. Suyama soon found out what had happened. The ambassador felt a slight cold. The solicitous bride had given him a Contac pill. If one is good, she reasoned, two would be better. Two knocked out the ambassador. I am glad to report that he soon felt better. The energetic young bride had a wonderful time cooking all sorts of delicate morsels for her husband and the clear climate of Maine helped to make for a delightful honeymoon.

The lodge is a beautiful log-cabin structure overlooking the bay and across to Bar Harbor. The main room is three times the size of a normal living room. I had had the Maine artists Stell and Shevis decorate the place and I must say they did an outstanding job. Of course, to begin with, we had all early American furniture. Stell made beautiful rugs and wall hangings in startling colors, using the seaweed and the starfish as motifs. Her husband, Shevis, had made a wooden mural over the fireplace, using the sun, waves, gulls, sailboats, and piers, all objects of Maine folklore, as sculptural artifacts. I must add that the fireplace is about fourteen feet wide; you could literally roast an ox in it. The twin-bedded room and adjacent bathroom are small, and so is the efficient kitchen nearby. We also keep two sleeping bags available, which can be used on an upstairs balcony. We tell the ambassadors that the place is convenient for four, six in a pinch. Across the road is a small one-room twin-bedded cottage with its own bathroom. Some couples come up with a chauffeur and maid. Some drive up alone with one or two children. Every once in a while an ambassador will come up with five or six children. Somehow they all seem to adjust and have a good time to boot. We explain beforehand there are no servants and the only service we can provide is taking care of the garbage and providing firewood for the hungry fireplace. I tried to simplify the problem of garbage by putting in a disposal unit but I wouldn't do it again. Many of the young girls who come from small villages in foreign countries as maids have never seen such a contraption and to them it is an object of fear and trepidation. In spite of the succinct directions posted, peach pits, melon rind, and even paper often defeat our efforts to make life more efficient. Usually after the first day, however, everybody becomes acclimated to this new part of the United States and, of course, this is what is hoped for. We wanted these representatives to realize that there was something more to the United States than New York City. They enjoyed the freedom and the privacy at the lodge. They enjoyed being able to go to the nearby supermarket and picking out their groceries leisurely. Of course they could come anytime to drop in on us, which they did. I put in a small library on each side of the fireplace and they enjoyed reading some of the books.

I remember when I first set up the lodge I consulted with the two nearby neighbors, the Schieffelins on one side and Mary Richards and Margie Spock, the sister of the famous Dr. Spock, on the other. Both neighbors were enthusiastic about the idea. I then called the Governor of Maine. Remember that it wasn't until a few years ago that black men could travel with relative ease. Some ambassadors had had bitter experiences traveling through some states. Frankly, I didn't know whether Maine was ready or not. Barely a week before, a Negro actress had been rebuffed in Kennebunkport. I didn't want to take any chances. But ready or not, someone had to start. I told the Governor that I didn't want any international incidents. He assured me there wouldn't be and, as a matter of fact, he sent one of his commissioners to see me and to help me. I sent out my invitations to some of the Ambassadors I knew and, sure enough, the first one to accept was my good friend Louis Ignacio-Pinto of Dahomey, now a member of the International Court at The Hague. The Governor's office and I wanted to make sure that everyone in Maine would know the Ambassador if he saw him.

As the Ambassador got off the plane, the television camera whirred, reporters rushed up, and the City Manager of Bangor sent a welcoming team. The Governor, who was opening a festival elsewhere, sent him a telegram. The plane had been late, but as we got into the car, I happened to turn on the 7:30 local news, and purely by chance timing, the newscaster was saying: "Tonight the Ambassador from Dahomey is arriving in Maine as a guest at the Fairleigh Dickinson University Lodge. . . . He will have dinner at the Le Domaine Restaurant."

When we got to Le Domaine, evidently a number of people had

come there to see the Ambassador, because the same newscast had been repeated every hour at the half hour all afternoon. It was an experience for us, too. Did you ever walk into a public restaurant where thirty or forty white guests are eating and you are with the only colored guest? Now it is no longer a problem. America has moved forward. But at that time I wasn't sure. I will never forget the experience of walking into the Oak Room of the Plaza with Ambassador Ben C. Limb of Korea. A drunken guest walked up to him and, mistaking him for a Japanese, mumbled some unfavorable reference to the war. One never knows! Meanwhile, the owner of Le Domaine gave the Ambassador top treatment and some ladies even came up and welcomed him to Maine.

As it happened, the next day was Open House Day in Bar Harbor, for the benefit of the local hospital. It was a splendid opportunity to see some of the outstanding homes and flower gardens. But everywhere the hostesses were delighted to see the Ambassador. Mrs. Edsel Ford, especially, was most gracious. Since the Ambassador's first language is French, the conversation most often broke into French and many of the dowagers were able to dust off their finishing-school French and engage in spirited, if somewhat ungrammatical, discussions.

On Sunday we attended the local community church. The minister, Margaret Henrichsen, made a point of welcoming him during the service. Many of the parishioners came up and were most cordial. I was happy to see this genuine reaction, because the Ambassador had had two unfortunate experiences in the United States, one in Maryland and the other in Knoxville. Perhaps this sojourn in Maine might help erase the bad impressions gained elsewhere.

Mr. Ignacio-Pinto is probably the most experienced and best trained of any of the African Ambassadors. He had been a Professor in Bordeaux, had a law office in Paris, and had been both a Senator in France and a Minister. While his background is thoroughly French, he has also had training in England. We talked endlessly about Africa and we all learned a great deal. Cocktail parties inevitably turned into parlor seminars.

Getting back to the Ambassador, one can never tell when an embarrassing turn may take place. At the home of Meyer Davis, the famous orchestra leader, one of his grandchildren asked Mr. Ignacio-Pinto quite innocently, "Why are you so black?" That same night, listening to the eleven o'clock news, the latest developments of the Georgia jailings cast a pall in the living room.

But all in all, it was an auspicious beginning for the University Lodge. The night before Ambassador Ignacio-Pinto left, he said, "Peter, for the first time I feel I've gotten to know the American people—je connais le peuple américain." And I understood what he meant, because, when he paid his protocol call on President Kennedy, on the way back he stopped at a hamburger stand in Maryland. At first he ate outside but, because it got windy, he decided to go indoors. He was stopped at the door and told that colored people were not allowed inside. To him this was inconceivable. Here he had just been welcomed at the White House, and now he could not eat in a public restaurant. To say that this was a traumatic experience would be putting it very mildly indeed.

As the ambassadors kept coming, we were extremely careful never to inject any propagandizing allusions into our conversation or in the library at their disposal. We pointed out how Maine had in many ways the same problems as a developing country: no great industries, except lumber; emphasis on tourism; reliance on small gardens, fish, and livestock. But the Maine people had something the big-city people did not have: a beautiful land, natural food from the earth and from the sea, absence of smog, freedom from keeping up with the Joneses. The neighbors were solicitous. Mrs. Whyte, when she learned that one of the visitors was a Scotsman, came over with scones only a fellow Scot could make. When one of the ambassadors drove his shiny new Cadillac off the road, Mr. Campbell laboriously built a crisscross of logs to raise it so that the car would not be bruised. Polly Richards, who grows all of her vegetables organically, would bring over vegetables from time to time.

After a while it was no longer necessary to make sure that everyone recognized the ambassadors and so we kept them incogniti, but in times of international stress some of the reporters from Bangor got through. When the Russians invaded Poland, one enterprising reporter figured that there might be an ambassador at the lodge. Sure enough, he ferreted out the Irish ambassador, Cornelius Cremin, and got a front-page story.

Many of the ambassadors became lecturers at Fairleigh Dickinson University. In many cases their compatriots studied at the University and in one case, Michèle, the daughter of Ambassador Tremblay, was a student at our Wroxton campus in England. We found the children of the various ambassadors almost without exception very bright and precocious and we encouraged nearby children to socialize with them. A number of ambassadors whose children were away from New York City found that the sojourn in Maine gave them an opportunity to be with the young ones. Most of the guests have become lifelong friends and have visited us in Rutherford. Ambassadorial life can be very rigid and studded with protocol. At times it can become almost cynical and artificial. But when these men were in East Sullivan, they felt that the free air of Maine once again made them human beings unfettered by the demands of their office.

In some cases, their own nation had come up so quickly that they were not familiar with some of the things we take for granted. I remember showing one ambassador my small, almost pocket-size, dictating machine. I explained how I would dictate into it whenever a thought came to me, no matter what city I was in, and how I would send the tape to my secretary. He was fascinated beyond words and I know that he adopted the system when he got back. They were fascinated with cars and, while in the city most of them for prestige reasons had chauffeur-driven limousines, it was in Maine that they could try themselves out driving a large car—not always with felicitous results because in two cases they managed to drive off the road into a culvert.

And we learned a lot from the ambassadors. We learned to have a greater feeling for African and pre-Columbian sculpture. We learned about foods, and we shall never forget how Ambassador Soo Young Lee of Korea brought along the beef to cook personally the first Korean meal we had ever had. Or how Ambassador Zenon Rossides of Cyprus prepared a natural-food luncheon that used the same ingredients known in the Mediterranean a thousand years ago. Later Zenon, Buckminster Fuller, and a group of us were to discuss the idea of a new country, dedicated to mankind and peace, and to be established on two hundred acres in Cyprus to be deeded by Bishop Makarios.

I had no telephone in the lodge, otherwise it would have been too much temptation to business as usual during their brief surcease. But emergency calls could always come through to us, in which case we would relay them within five minutes. And they did come through, reminding us only too vividly that painful events do take place in spite of the peaceful atmosphere at the lodge. Once in a while a call would come through in the early morning or late at night and every so often a delegate had to rush back, as did Ambassador Cremin during the Czechoslovakian crisis, and Ambassador Rossides because of an emergency session of the Security Council.

My interest in international activities has naturally involved me in many interesting experiences, as many other university presidents have been. I ran my third conference in Seoul, Korea, as president of the International Association of University Presidents and this had some curiosities. I wanted the conference to be an opportunity for the many administrators in Southeast Asia to discuss higher education. The official speakers were to speak briefly and the plan was to have as much discussion as possible from the floor. I had Professor René Dumont of the University of Paris come and speak on education in an emerging nation. He had been a consultant to a number of such nations and knew firsthand of the many large-scale errors that are made in a new country. And he had a vivid understanding of the role educators can play in leading a new country. Another friend, General Carlos P. Romulo, who as Chancellor of the University of the Philippines understood the problems of an expanding university and had used his experiences in American universities in planning a new role for his institution. Last, I had His Excellency Thanat Khoman of Thailand, who not only was a leading statesman in Asia but had gone through the labor pains of establishing a new university in his own country.

I must say that the entertainment arrangements for the three-day meeting were the most extravagant I have seen anywhere—so extravagant that they completely overshadowed the educational aspects. Triumphal arches all over the city welcomed the conference. The government issued a special stamp for the occasion. The President of Korea had a special reception, the Vice-President a dinner, for all those attending. The Mayor of Seoul, the Foreign Minister, the Chairman of the leading political party, and a number of others all gave parties. Chairman Jong Pil Kim of the Democratic Party gave a full musical extravaganza and dinner dance for those attending.

The conference hall was almost like a United Nations' Assembly hall. Each delegate had a nameplate made of teakwood, with his name in mother-of-pearl carved in it. The only trouble was that, as is bound to happen, many delegates never got there, and many who came at the last minute had no nameplates. There was super organization from the local end and it was all very impressive but, believe it or not, I could not get an exact list of those in attendance. There were scores of willing and smiling young ladies ready to say "yes" to anything. We soon found out that this was a mechanical answer and in practically every case the young lady did not understand the question and said "yes" in order not to offend the person. Finally, I sent an assistant around personally to each desk to get the exact information I wanted.

Then, as a matter of courtesy, I had the meetings chaired by the local chairman. What I didn't know is that a number of the college presidents in Korea, excited by the idea of an international academic conference—the first of its kind in Korea—had had their comments on higher education printed in pamphlet form. Then a few timidly asked to present a "few" comments. Permission was granted. The few comments turned out to be forty-minute speeches delivered in a flat, unaccented, uninflected mechanical monotone. The dikes were open. Soon every other college president in Korea wanted to expound his ideas. What to do? My first impulse was to rule out all nonscheduled speeches and to limit any remarks from the floor to two or three minutes. I had followed this procedure at the Wroxton conference and it worked very well. I had a hurried conference with General Romulo, Dr. Thanat Khoman, and Ambassador Ben C. Limb. They counseled that if I ran roughshod over the well-meaning Korean educators, we could be accused of American lack of appreciation for the Asian way of doing things. It sounded like a high-level diplomatic conference. I then checked with an ad hoc group of American presidents. They felt the same way. It wasn't until General Romulo, Dr. Khoman, and Dr. Dumont spoke that the speeches could be heard and understood.

As a sequel to the international meeting, I led a group of college and university presidents, all members of the Association, on a visit to Taiwan, the Philippines, Thailand, and Sikkim. We were feted royally in all places. In Taiwan, both His Excellency Mr. Liu Chin, Chinese Ambassador to the United Nations and Major General Richard Cicolella, United States Military Advisor, had paved the way so that most of the Ministers of State were on hand to greet us at the airport. Throughout our stay we were treated as state guests and had full police escort everywhere. We visited the leading universities, where warm welcoming receptions awaited us. The Vice-President gave a fifteen-course dinner in our honor that was a culinary tour de force.

There was a reception given by Generalissimo and Madame Chiang Kai-shek. Each president had an opportunity to speak in turn to the Generalissimo through an interpreter. Each wife had an equal opportunity to speak with Madame Chang Kai-shek but, since she speaks English faultlessly, no interpreter was needed. Major General Cicolella's party was exceptionally fine, with the Generalissimo's younger son as an unexpected guest. The highlight was a flight to Quemoy. It is an unwritten understanding between Communist China and Taiwan that each may shell the other on alternate days, although this is rarely effected. Our flight was on the day the Communists would not shell, but as an added precaution we had military protection, just in case. The reception on the island was enthusiastic. To see an island where the whole population can live underground if necessary is an awesome experience. Chang Kai-shek's older son, General Chiang Ching-kuo, who was almost shot on his visit to New York, offered another gourmet lunch, this time in an underground cave.

In the Philippines, the group learned a great deal about the proprietary universities, which fulfill a temporary role until they assume the nonprofit role of American universities. Senator Anita Benitez, whom I had known, had an outdoor fiesta on her beautiful estate, with President Marcos as the guest of honor.

In Thailand, Thanat Khoman, now the Foreign Minister and whom I had known for years when he was a young diplomat at the United Nations, took us in tow. It started with a Thai dinner in a quaint restaurant. We had the same receptions as were held for President Johnson when as Vice-President he visited Thailand. Premier Kittakakorn held a dazzling dinner and show for us. It was a fairyland scene. The King and Queen received us and for an hour and a half we sat around and discussed in seminar fashion the problems of the younger generation. Finally, there was a party by the heads of Bangkok College, an institution that two of our deans had helped to organize. In between were many visits to educational institutions and we began to see the tremendous problems of financing, of library needs, of faculty preparation, of lack of laboratories. Another long-time friend, Prince Wan, whom I had known when he was President of the United Nations' Assembly, was most considerate. As Chancellor of Thammasat University, he arranged what was really a seminar on higher education.

And then to the top of the world, Sikkim, where Hope Cook, the American girl, had married the Prince who now is King. After a hair-raising, constant, switchbacking ride for six hours over roads that were often washed out, we reached Gangtok, nestled in the Himalayas. We stayed at the Royal guest house, which was stocked with cold orange juice and all sorts of liquors. The reception by the King and Queen was the highlight of the visit. We saw the microcosm that was Sikkim, and the valiant efforts to create schools and cottage industries. And throughout, as in the other countries, we acquired a better understanding of Asian religions and their role in the order of things.

In the early part of 1966, I suddenly felt that I could not for long continue to work as hard as I did. I further felt that it was unfair for my wife to work so much, for we were devoting eighty to ninety hours a week. But the number of hours was my least worry. I was sixty-two, and while I apparently enjoyed good health, as a complete clinical check-up showed, I did become preoccupied with my health and obsessed with the idea that I might begin to flag. Perhaps the time had come for a younger person to take charge. I worried too much. After all, there were four campuses in New Jersey—Rutherford, Madison, Edward Williams College in Hackensack, and the newest, Teaneck. There was also Wroxton in England and a Marine Biology Laboratory in St. Croix in the offing. As the university emerged over the years from an old castle in Rutherford, I had been involved in all of the details, whether faculty, curricular, financial, or day-to-day mechanics. It seemed to me at the end of the day that there were always more problems that lay ahead. But I suppose there was a feeling that my kind of a Quaker-type of academic community was at an end. As I have stated before, we had at the beginning of our institution a simple organization whereby faculty and students were involved in everything that took place. Now everything was done through ever-so-many committees, and the fact that we were a multicampus university made things much more cumbersome and time-taking. I used to take a great deal of pleasure in dealing with students directly. We had them at our home. We used to go to all of their functions. We knew their parents. We would often discuss them with high school personnel. We would follow closely their progress after graduation. It was like one big family. Now most of that was behind us. We still spoke to students, but the golden thread had been broken. We were now too busy to feel the glow of that person-to-person relationship. I decided to sleep on the matter until September and if at the end of the summer I felt the same way, I would ask the Trustees to seek a new president. In September I still felt the same way and I then decided to have a talk with the chairman of the board, Ed Williams. I went to see Ed and told him that I felt the time had come to seek a new president. I stated that it would take a year to find one and that I would be glad to pass over to him everything I knew. Dick Dickinson happened to be in his nearby office. Ed called him in and relayed my message. Dick said, "Why don't we get you an assistant and if he works out, when you reach 65, he can take over?" I said, "No, let's give him the full dignity of the office and the powers that go with it." I added that I would be glad to work out the transition in any way the trustees saw fit, without any strings attached. Dick then suggested that I become Chancellor, and Ed quickly agreed. The details would be worked out later.

I left the meeting with a great sense of relief. Was this part of growing old? Was the desire to avoid heavy responsibilities part of the process of aging? I had had twenty-five years of close contact with every little detail and I was tired of it. But as I left the offices of Becton, Dickinson and Company, where both Ed and Dick had their offices, I felt a renewed sense of energy. I could now work my head off but it wouldn't be forever. One saw light in the offing.

But, in spite of my feeling that the time had come for me to retire from the endless tasks of day and night, twelve months of the year, I wanted to make sure that a doctoral program would be established in the College of Education. I had been able to have the old Board of Education in New Jersey approve such a program. In the meantime I gave instructions to the librarians at Rutherford and Teaneck to build into their libraries the books and publications needed for such a program and, in fact, contributed money for this purpose. In my application to the State, I stressed the importance of independent research, of knowledge of the community, and an overall need for school leaders to be innovative and creative. In our own plans I established further guidelines. I have seen how so many candidates for doctoral degrees have frittered away years of desultory study, sometimes because of their dilatory approach but very often because of lackadaisical methods of guidance by the university involved. To me it is just short of criminal to let a graduate student take course after course and then to discourage him at the last barrier by prolonging the thesis-writing, by changing advisors, or by simple neglect. I proposed that we be very careful in our selection of new candidates. First, let us limit them to those who have their master's degree and have mastered the skills of reading, writing, and speaking. Sounds silly, doesn't it? Yet I have seen so many would-be doctoral candidates lacking these prime essentials. The goal should be to train leaders in education, not mousy men or women who might be good in digging out little-known facts in some esoteric field but who wouldn't know what to do in a tinderbox school or community situation. We wanted candidates who could dig out facts, who could organize material and complete a meaningful thesis study. But we also wanted to train dynamic men and women who could rise above the general mass and point the way to greater service to his or her fellow beings. Once having chosen the candidates, then I felt that they should be brought in, with their spouses if they were married, and made to realize that they had to concentrate on the doctoral studies for three years, twelve months of the year. There should be no distracting activities, but they could work at their jobs. Their social activities for this period, sports, vacations, and other professional pursuits would have to be put aside. They would be expected to put in seven hours a day, seven days a week, 365 days a year. Further, I felt that the whole group should be brought together once a week, with their spouses, incidentally to participate in important discussions led by an outstanding leader on the persistent problems of living. Last, I felt that each candidate should have a weekly conference on his reading and on his research. A consistent program such as this, week in and week out, can't help but produce outstanding graduates. It took a few years, but I was glad to see the program get under way in 1971 and I predict that in time the leaders in education, especially in the northern part of New Jersey, will come from this group.

But fate has a strange way of altering human plans.

It was in the late spring of 1968 that I received a call from Mr. Huntington Hartford inviting me to lunch at the River Club in New York. I happened to be free that day. Through his friend Sandy Williams, son of our very good friend Peggy Williams, former publisher of the Paterson Call, I learned the story later. Sandy had come home to Paterson, which home he shares with his mother. He had told her that Mr. Hartford, having reached the age of 57, felt that the time had come to put the Gallery of Modern Art under a broader sponsorship, probably under the aegis of a university. Peggy had immediately suggested that Mr. Hartford contact me. The luncheon led to a few other meetings, one with Ted Sorenson, his lawyer. The financial requirements were out of our range and I dropped the matter. Some time later, while I was at the Hilton in London, I received a transatlantic call at 1:30 A.M. from Mr. Hartford, who asked me whether I was still interested. I told him we were, but that financially we couldn't handle it. He told me that some arrangement could be worked out. When I returned to the United States, I asked Ned Feldman, an international banker and a member of the University's Board of Fellows, to look at the proposition from a very hard-boiled point of view to see whether it made sense for us to consider it. He studied the new proposal and reported that we should by all means try to reach an agreement. Financially, the University could not lose out; from the point of view of prestige, it would be a great achievement. After interminable meetings which Mr. Samuel J. Silberman, a Trustee, and Mr. Feldman spearheaded, the final agreement was reached.

Having now retired from the presidency of Fairleigh Dickinson University, I, with Sally, was looking forward to a lazy year and we had planned three trips. The first was to retrace the probable course of Ulysses in the Mediterranean. The second was to Southeast Asia, where a number of my friends wanted advice on their colleges. The third was to West Africa, where the interest in my help came from the official government sources. As the possibilities for the acquisition of the Gallery became more realistic, we all felt that it would make sense for Sally and me to get into the workings of the building so that we could measure the possibilities and also be ready for any transition if it occurred. It was a complicated procedure. Mr. Hartford's personal fortune, the foundation he had established, and our university had to bring their interests into agreement. An official ruling from the State Supreme Court was necessary for Mr. Hartford to reserve half of his personal trust, payable to the University upon his death. Since this trust is now worth $8,000,000, the half going to the University would be $4,000,000. However, since Mr. Hartford was 58, assuming normal expectancy of life, this fund would probably double at least twice and be worth eventually $16,000,000. There was a mortgage of $3,500,000. He paid off $1,000,000; Mrs. Fairleigh S. Dickinson Sr. paid off $2,500,000. Mr. Hartford also agreed to meet a substantial part of any deficit as it occurred.

In the meantime, we did not know from one month to another whether the transition would ever take place. While Sally and I occupied offices in the building, we really had no power. I did not dare to hire any personnel. Activities at the Gallery had practically come to a halt and there was the barest of a skeleton staff. Somehow I managed to put together a series of exhibits, drawing upon my experiences at the University.

During the summer of 1969, we went to Maine feeling that nothing would happen for a long time. But four times I had to rush back because of lawyers' conferences. Finally, on August 12, the project was completed. On August 14 we had a press conference, and both Senator Dickinson and Mr. Hartford were on hand. As is always the case, there is a tussle between the New Jersey dailies and the *New York Times* for a scoop. The *Times* had a front-page story, with a picture and continuation on an inside page. For those not cognizant of the tremendous importance of the *Times* all over the country and throughout the world, let me say that this was one of the greatest accolades that could come to any University. To share the front page of the *New York Times* with the most important happenings all over the world is a prize that any public relations director would give anything to effect. In our case it just fell into our laps. Grace Glueck of the *Times* did a beautiful job. She sniffed the story beforehand and was quietly researching it. On the day it happened she called me immediately to check on details. I realized that she was a stickler for accuracy. I stayed glued to the telephone in case she called again. Sure enough, she called back four times to make sure of some minor point. Her final story was the most professional job I have ever seen. Guy Savino of the *Newark News* somehow had got wind of the transition in the Gallery. He called me to ask me what I was doing in an office in the Gallery. I couldn't give him any information but he was enough of a newspaper man to run an imagined story a day ahead of time, even though there were no details. The *Record* in Bergen County, the newspaper nearest the University, got into the act on a split-second change. All of the New Jersey papers, as well as hundreds throughout the country, finally came out with the complete and correct story.

The new name decided upon by the negotiating committees was "The New York Cultural Center in association with Fairleigh Dickinson University." The Center is an autonomous body incorporated in the State of New York. There is a cooperative arrangement with the University but it may not cause any expense to the University. The Center functions as an alumni club for Fairleigh Dickinson graduates and also as a Faculty Center. It is easily reached from Bergen County across the Hudson River, and its proximity to the theater section and Lincoln Center makes it a convenient meeting place for the Fairleigh Dickinson family. Various groups in Bergen County, including the two Town and Gown Societies, have various functions at the Center. To those who know the juxtaposition of the two counties and the mechanics of travel, the advantages are only too evident.

Once the acquisition of the Gallery was a fact, we began to seek a director, and after a few months were fortunate in choosing Donald Karshan, who had organized the Museum of Graphic Arts. As a matter of fact, some six months before, I had arranged for the showing of his beautiful print collection, which is one of the best in the world. In discussing matters with him, both Sally and I had been impressed with his professional knowledge, his organizing ability, and his energy. For those unfamiliar with the museum field, let me point out that a museum may show part of its permanent collection and also arrange for temporary exhibits which come in from the outside. In our case we had no permanent collection, although many persons mistakenly assume that we inherited Mr. Hartford's personal collection. Our entire art activity consists of arranging for loan exhibits, some of which we originate and may then loan to other museums, and some which may come from other museums, from other countries, or from individuals. In Manhattan, we are competitive with the Metropolitan, with the Museum of Modern Art, with the Guggenheim, and with the Whitney. More specifically, we are competing for the attention of the most sophisticated audience of the world, the people and the critics of New York City.

Sally and I, as full-time volunteers, were plunged into this intensely competitive milieu. During the interim period, my very good friend Alfredo Valente, although recovering from a heart attack, valiantly joined us as a volunteer curator until Donald Karshan took over.

Having been born in New York and having received most of my education there, I felt that in a sense I was helping to pay back what a great city had given to me. There is no question in my mind that New York City, like most cities, is retrogressing. The only way to arrest this tendency is to create or maintain islands of beauty. As I look at Manhattan, it seems to me that Columbus Circle at the present time is the hub of that city, with one arm running along Central Park South to Fifth Avenue and another arm north to Lincoln Center. In the past the hubs have been in the southern end of the island, at first Canal Street, then probably Fourteenth Street, then Twenty-third Street, then Herald Square, then Times Square. Each in turn has retrogressed, has melted into a nondescript second-rate commercial area and, in some cases, has even become honky-tonk. If the triangle I have described above deteriorates, then New York City is finished. It can't go further north because further north has already been built up and is already deteriorating. I take pride in what we are doing because we are helping to keep up the beauty of the city. And I believe that the city owes a real debt to Huntington Hartford for erecting the gallery that is now the New York Cultural Center at this vital spot in Manhattan.

I find the greatest enjoyment in speaking to the pupils from the elementary schools when they come in groups to visit the Center. I remember the many classes that came to see the exhibit of photographic reproductions of Michelangelo's Sistine Chapel ceiling. Originally made at great expense by the Capitol Broadcasting Company who, with the permission of the Vatican, had constructed a 65-foot scaffolding in order to photograph the ceiling at close range, the exhibit showed the true colors of the ceiling as they had never been seen before. It was Huntington Hartford's idea to bring the photographs into the Center and, as a matter of fact, he spent about $25,000 to have the enlargements made from the transparencies jealously guarded in the temperature-controlled vault of the Capitol Broadcasting Company. Mr. Milton A. Fruchtman of Capitol—who had enthusiastically initiated the Vatican proceedings and lovingly nurtured its various outgrowths, including a stupendous motion picture—worked with us hand in hand.

Alexander Eliot, former art editor of *Time* magazine, told me that he and his wife, Jane, had read just about everything that Michelangelo had probably read before doing the ceiling. When a visitor goes to the chapel he sees the ceiling through 65 feet of distorted light and dust. The Capitol Broadcasting people had had the accumulated dust of centuries taken off to reveal the real colors that Michelangelo had painted on. Alex said it was the most intense emotional experience of their lives when they would lie on their backs on the platform of the scaffolding and look at the ceiling for hours.

At any rate, I would take the young people around to the different photographs and eventually we would get to the one showing God creating Adam. There is nothing more soul-filling than to be talking to a group of children seated on the floor and to have their complete attention and participation. I remember one youngster's asking

"How did Michelangelo know what God looked like?"

I explained that this was what he imagined him to look like and that each person has the right to imagine God in his own way.

"Could he have been black?"

I told them that the writer, Marc Connelly, had portrayed God as a black man and that the angels were all black too.

One girl piped up and said "I think God is a woman."

"Naw," answered another lad, "a woman wouldn't be strong enough to be God."

These were young people of ten or so. Later, one youngster wrote to tell me how much he had enjoyed the visit and added that I would make a good guide at the Sistine Chapel.

On another occasion I was taking a group through a photographic exhibit of Women Throughout the World. It was directly in line with the Women's Liberation movement and, even though the children averaged about eight or so, they participated enthusiastically and intelligently in the discussion. As we were passing one corner of the Gallery, one child burst out "What's that?" I looked and realized it was a photograph of a baby coming out of the mother's womb. I suddenly realized that I had a sex-education project on my hands. I made a quick decision and said in my most nonchalant tone, "Oh, that's how babies are born. That's a baby coming out of its mother." I added a few pious words on how much mothers have to suffer during childbirth. I think they were impressed. Later I received a letter from one child saying that when he got home he told his mother how babies are born.

Then we stopped in front of a photograph showing women washing clothes in a small stream in one of the emerging countries. I explained to them I had seen scores of such streams where not only do the women wash their clothes but people bathe in the same stream and use it as a toilet. But to top it all, they also drink the water. They all grimaced at the story and one boy even took umbrage at my use of the word toilet. "Couln't you use a better word?" he asked. I turned helplessly to one of the teachers accompanying the group. Here I was, trying to use a word I felt they could all understand, and it backfired on me. But most of the children felt it was awful that they drank the dirty water and that the children played in what they considered an unsafe stream. This was made to order for a motivated discussion on pollution. I told them we were, in effect, doing the same thing in America, only on a much larger scale. I explained to them how our toilet and factory waste goes into all of the large rivers, how beaches are polluted, how the water we get from many rivers has to be chlorinated. I asked how many had noticed the awful taste of water in New York City. Most hands went up. Since most of the children came from ghetto areas, I think it began to sink in that our problems were just as serious. I went further and told them about a little lake in Maine. Toilets emptied in it, people swam in it, and the water from it was used for washing machines and for drinking. Wasn't it the same thing as the stream we were looking at? I told them I bought bottled spring water because I couldn't stand the taste of water in my town in New Jersey. I know they learned something, from the letters I got from many of them.

When I had the sculptures of Jacob Epstein at the Center, I invited blind adults and children to come and feel the works of art. This is the only kind of art blind people can appreciate. With the help of The Lighthouse, I had a catalogue in Braille giving the description of each piece. On each sculpture I had a label, also in Braille. I had already done this at Fairleigh Dickinson University, where I established at our Teaneck campus a small permanent collection of copies of many important small sculptures from a number of museums. I had been able to do this with the help of a friend, Mrs. Leo Pollak, and it had worked out very successfully for a number of years.

But we all find innumerable opportunities to serve people and this is what lends meaning to life. I maintain that there are only four things that make life worthwhile: 1) serving other people, 2) communicating with people, 3) feeling exhilaration at beauty, whether created by God or man, 4) feeling able to create. Everything else stems from these four basic things. But to be able to function in these four areas, we need an élan vital which in turn is dependent on good physical and mental health.

On one of my inspection tours through the gallery, I saw a disheveled young man. He hardly looked like the type that would want to spend a dollar to come in and look at works of art. I got into a conversation with him and he spoke very intelligently about aesthetic matters. I asked him what he did. He said he was a composer of piano music. I brought him up to our eighth-floor lounge, where there is a grand piano, and asked him to play one or two of his compositions. I was somewhat dismayed when he took out a bunch of keys, picked up some cutlery, a tray, and a glass from a nearby serving stand, raised the piano top, and began to play some of the most unusual music I have ever heard. By rolling the side of the glass over the strings with his right hand and accompanying with his left, he produced some eerie effects. He then jangled the bunch of keys on the strings and a different effect was produced. He used various pieces of silverware to produce even stranger sounds. By this time I felt I owed an explanation to the people sipping drinks in the lounge. I told them frankly what it was all about. We all listened, somewhat perplexed, but the music wasn't unpleasant and when he was through we all burst forth into genuine applause. I invited the young man, Burton Greene, to give a series of concerts at the Center and invited a coterie of his friends to them. It was an exposure for him and I hope we gave him a little push in the career he had carved out for himself. At some of his concerts, a friend of his, Margaret Beals, a dancer, improvised to his music. There were no rehearsals. In some cases Burton had written out some indication of the music with stretches to be filled in by the pianist. In others, he improvised as he went along. But Margaret improvised completely, and I must say she was the most fascinating dancer I have seen. Very good looking, there wasn't a false note in her movement or in her body. She would leave the stage, go down the aisle, out one door, in another, while Burton was improvising away. I know it all sounds odd, but we all enjoyed the performances. There was a spirit, a sincerity, a vibrant quality that only young people can have, and, after all, this is one example of beauty created jointly by God and by man.

At the University, many years ago, I decided to form a club for persons over 65 years of age. I asked the heads of churches to suggest names of people to invite. It was Sally who suggested the name for the group: Allegro Club. They met every Friday and we all took turns speaking on some subject to the group. I also invited them to use our cafeteria and library, and to feel free to roam around on the campus whenever they felt inclined. They also came to many of our special events, such as lectures, dramatic presentations, films, and concerts. It meant that these persons, most of whom were alone, suddenly were members of an interesting social group. It also meant that they had all the advantages of a university club, so to speak. When we acquired the gallery, this was one of the projects on the agenda. I find that people in New York are very lonely indeed. What can make more sense than to be a member of such a group at the New York Cultural Center? Furthermore, it's good to have a club to walk to or even take a subway ride to. After all, what can you do at a club? Meet friends, have a meal. This you can do at the Center and, in addition, have art exhibits to see, films to watch, and good music to listen to. Service to people, communication with people—these are the things that mean life. Whether at the University or at the Center, these are the things that have enriched my life and Sally's.

When the Allegro Club in Rutherford was fifteen years old, I was asked to be the anniversary speaker. As I began to talk to them, I realized suddenly that I too was now a senior citizen. As the years creep up on you after sixty, it's always the person five years older than you who is old. In your own mind you are still young in spite of creaking joints and a slower gait. I would say that most people, including us, are about five years behind in catching up with real age. And so, as we try to catch up and as the years roll by, Sally and I try to give our time and energy to helping causes, institutions, and people. We are thankful that we are given the opportunity to serve and that through the years we may have acquired some knowledge that will be helpful to others. While we still enjoy crowds, we find greater enjoyment in small groups—six or eight people.

When President Nixon appointed me to the Board of Foreign Scholarships, it gave greater scope to my leisure activities. I had had a taste of such activity when President Johnson had appointed me to the Advisory Board for the Peace Corps. Now I was involved again in the sort of activity on the international scene that Sally and I enjoy so much. By pure chance I was given an assignment in Africa just as I was organizing my fourth international conference, this time in

Monrovia, Liberia. Although I knew quite a bit about Africa and had been there a number of times, we wanted to know more and plunged into a small mountain of books and reports. But this is what lends zest to life—the exploring of new fields, an opportunity to serve a new group of people, the excitement of change, the making of new friends. And, as we have moved from the aggravations of daily operation to less frustrating but still challenging activities on a different plane, I still think wryly of that corny motto which I adopted on the spur of the moment: "Fortiter et Suaviter"—Bravely and Pleasurably.